False Messengers

False Messengers

How Addictive Drugs Change the Brain

David P. Friedman, PhD

Wake Forest University
School of Medicine
Winston-Salem, North Carolina, USA

and

Sue Rusche

National Families in Action
Atlanta, Georgia, USA

First published 1999 by Harwood Academic Publishers

Reprinted 2002
by Routledge, 11 New Fetter Lane, London EC4P 4EE

Routledge is an imprint of the Taylor & Francis Group

Front cover: Neurotransmitter molecules (green horseshoes) seek out receptors (purple tubes) protruding from the cell membrane (peach surface) of a neuron into a synapse. Pentagons in the lower part of the illustration represent second-messenger molecules generated inside the neuron in response to the binding of the transmitters to the receptors. Illustration by Gayle Gross de Nuñez and Rochelle D. Schwartz-Bloom, SAVANTES, Durham, North Carolina, USA.

Illustrations throughout the book by Eric Bowles and Robyn Connell.

Printed and bound in Great Britain by TJI Digital, Padstow, Cornwall

---

British Library Cataloguing in Publication Data

Friedman, David P.
    False messengers : how addictive drugs change the brain
    1. Brain—Effect of drugs on
    I. Title   II. Rusche, Susan
    615.7′8

ISBN   90-5702-5159

To our families

Susan Friedman
Zachary and *Max Friedman*

and

Harry Rusche
Philip, Marian and *Alexander Rusche*
Steven Rusche and *Kristen Jester*

CONTENTS

FOREWORD

Substance abuse remains one of the major public health problems affecting the United States today. Both the numbers of people addicted and the damage done to them and society are staggering. Up to 50 million people are addicted to nicotine, about 12 million to alcohol, over 2 million to cocaine, and almost 1 million to heroin. Deaths from these addictions amount to more than 500,000 per year. Remarkably, even though the dangers of addiction are usually well known, people still initiate drug use and continue using drugs until they become addicted. Only then do they discover just how difficult it is to stop using drugs, even after they have encountered the many problems that result from such use. People lose their jobs, health and families, and still continue their drug use. Why addicts continue drug use in the face of all these problems is a source of both amazement and pain to their families and friends.

False Messengers provides an exciting look at why neither knowledge of the problems that drugs can cause nor actually suffering their consequences is enough to change behavior. But there is good news: Knowledge about why addiction occurs is increasing. As it does, the hope that we will develop better ways to prevent and treat addiction will increase as well.

Addictive drugs mimic the actions of naturally occurring brain chemicals (hence, the title *False Messengers*), but do so with a speed and intensity that cannot occur in day-to-day reality. For many people, the intense pleasure that drugs can invoke is worth any possible risk—which people are good at denying, minimizing or believing will not happen to them, anyway. As use continues, changes that take place in the brain act to perpetuate drug use until it becomes the compulsive drug-seeking behavior we label "addiction." Both the drug itself and the circumstances of use affect and change the brain, often requiring treatment to help the person resist continued, destructive use.

False Messengers also provides insight as to why patients who take drugs for pain relief rarely become addicted, while the same amount of drug taken to get high often leads to compulsive use.

The psychological context in which drugs are taken can produce lasting neurochemical changes, as can the drugs themselves. A better understanding of this context could improve our current medical system, in which patients often get inadequate pain relief because of "opiate phobia" on the part of doctors and nurses who fear they will make patients into addicts.

Newspapers and television frequently portray stories of well-known entertainers, athletes and executives who jeopardize or throw away careers worth millions to continue their drug and alcohol use. Millions of unpublicized ordinary citizens lose jobs or end up in jail because they continue their destructive chemical careers. This book explains in terms the nonscientist can understand why this happens. It provides clues and tools to those who try to prevent drug use or get someone they care about into treatment.

It is often said that drug use is a preventable behavior and addiction is a treatable disorder. But it is increasingly clear that addiction erodes a person's ability to control behavior. Therefore, denial of the existence of a problem, as well as reluctance to enter treatment, is often part of the chemical picture. This explains why involuntary treatment—whether under pressure from family, employer or the criminal justice system—may be needed and can lead to results equally as good as those from voluntary treatment. The seductive "highs" of these "false messengers" can be so compelling that the person wants no part of having to give them up. The cognitive part of the brain recognizes the need to do so, but the emotional part denies such logic or need and recognizes only the desire to achieve more of the good feelings.

Readers of this book will learn much about our increasing knowledge of addiction and, in so doing, enrich their own lives and understanding of this problem that is such a plague to modern society.

Herbert D. Kleber, MD

Professor of Psychiatry and director—
Division of Substance Abuse, Columbia University
executive vice president and medical director—
National Center on Addiction and Substance Abuse

PREFACE

Listen to how some people describe what drug addiction is doing to someone they love:

My boyfriend is a cocaine addict. He is trying to quit, but every time he gives it up for a while, a temptation hits and he does it again. It is killing me not to be able to do anything. I don't want anything bad to happen to him.

My 40-year-old son has been addicted to heroin since he was 15. He has tried everything—cold turkey, Narcotics Anonymous, many treatment centers, prayer, methadone, and on and on. I feel as if I am watching him die.

We recently found out that our 16-year-old daughter has been on marijuana for about a year. She was always a good student, but now her grades are dropping. She has been a loving girl, very involved with her family. Now she doesn't come home at night until very late. Sometimes she doesn't come home at all, and we are terrified something has happened to her. She rarely does what we ask her to do anymore.

I am a 40-year-old nurse with a 63-year-old mother who is an alcoholic. She was hospitalized with short-term memory loss last week, after taking a tranquilizer, drinking two beers before work and eating no breakfast. Today, I left my job 2 hours early and drove 25 miles to get to her, because she couldn't remember her name or even what day it was. She resented my coming and lied to the doctor about her alcohol use, which begins every day before work. My father died from alcoholism. My two brothers are "weekend alcoholics" who play Mother's games. I do not drink and feel totally alienated from my family. I want to fire my mother from my life forever. I don't want to care any more.

> *I am 9 years old. My parents are taking drugs. I have been taken away from them once. I want to get my parents help, but I am afraid to tell anyone. I'm afraid I will be taken away from them again. I don't know what to do.*

These are excerpts from a few of the thousands of e-mail messages people send to a column on the Internet that answers questions about drug abuse and addiction. Millions of Americans, like the nine described above, are addicted to alcohol, cigarettes, cocaine, heroin, marijuana or other drugs. Each one began by freely choosing to use the drug to which he or she is now addicted. Not one started out saying: "I think I'll take this drug so I can become an addict."

Yet all addicts will tell you that they've lost their power of choice. They can no longer choose to stop using the drug they are addicted to, even though not stopping may mean losing a girlfriend, losing their life to addiction, losing the respect and trust of their parents, losing the respect and support of their grown child, or even losing their child, to cite the potential consequences for the addicts described above.

Not only are addicts unable to stop using drugs, but their behavior is dominated by the need to get and keep taking the drug to which they are addicted. This behavior—seeking and consuming drugs—takes precedence over all other behaviors in which they might otherwise engage. Because addicts' need for drugs is so great, it drives them to behave in ways and do things they never would, if they were not addicted. In essence, addicts don't just lose their ability to choose whether or not to use drugs, but because drug use so dominates their lives, *they virtually lose their free will.*

How can a little juice scraped from a poppy, or a small bit of powder processed from the leaves of a bush, or smoke from two different kinds of weeds, or the products of fermented grains have so much power that they can ultimately rob people of their free will?

While the question may seem astonishing, the answer is truly astounding. *Drugs of abuse mimic the natural chemicals the brain uses to communicate and carry out all of its functions.* No matter how they get there—smoking, injecting, ingesting or inhaling—once drugs enter the body, the bloodstream quickly carries them to the brain. And once they reach the brain, masquerading as the brain's natural chemical messengers, drugs send out false messages: intensifying some of the brain's messages, distorting others, completely falsifying still others.

Drugs *are* false messengers. Precisely how they act on and change the brain, change behavior, produce addiction, and ultimately steal free will—how drugs produce the behaviors just described—is the subject of this book.

ACKNOWLEDGMENTS

A number of people have made the publication of this book possible. We express our appreciation to our editor Sally Cheney, for her guidance and patience throughout the creation and completion of the manuscript, and to our copyeditor Alison Kelley, for her insights and many helpful questions that enabled us to clarify key points.

We also thank Dr. Elaine Johnson, former director, and Kent Augustson, former associate director, of the U.S. Center on Substance Abuse Prevention, which funded a meeting of scientists to advise us on the development of the book's content. We are especially grateful for the advice and counsel of those scientists who attended that meeting. The Scientific Advisory Committee consisted of Henri Begleiter, MD, PhD; Floyd Bloom, PhD; Larry D. Byrd, PhD; Steven R. Childers, PhD; Marian W. Fischman, PhD; John A. Harvey, MD; Jack Henningfield, PhD; Steven J. Henriksen, PhD; Judy Howard, MD; Allyn C. Howlett, PhD; Chris E. Johanson, PhD; Herbert Kleber, MD; Mary Jeanne Kreek, MD; Michael J. Kuhar, PhD; A. Tom McLellan, PhD; Roger Meyer, MD; Charles P. O'Brien, MD, PhD; and Robert Wilkerson, PhD. A special note of thanks to Dr. Howlett, who led us to our publisher.

We are also grateful to Paula Kemp, who helped write the *You Have the Right to Know* curriculum, and the parents and children of Bankhead Courts, Techwood Homes, and Usher Middle School in Atlanta, who participated in the "Right to Know" projects. These projects first led us to break down complex information about brain function into terms simple enough for parents and educators to teach children. Their enthusiastic response to this effort encouraged us to write this book.

Individually, Dr. Friedman thanks all the scientists, teachers, treatment providers, students and others whose intellectual curiosity and desire to understand what drugs do to us helped him develop his ideas about drugs and the brain. Similarly, Mrs. Rusche thanks all the parents, children, educators, health professionals, policy makers, law enforcement officials and others who have asked for help in the effort to prevent drug abuse and addiction. Interactions with people who want to know precisely how addiction develops have challenged both of us and helped us learn how to present this information in ways people can understand.

Finally, our thanks to our families for their patience and forbearance during the writing process.

CHAPTER 1

YOUR BRAIN, YOUR MIND —
YOUR CHOICE

*H*ave you ever thought about the relationship between your brain and your mind? Or how free will is related to either the brain or the mind? Or whether something that affects your brain can change your mind or alter your ability to choose what to do? Most people haven't. To understand how drugs of abuse cause addiction, all these concepts become important. Before we immerse ourselves in our study of drugs, let's briefly explore these fundamental concepts.

The Brain: The portion of the central nervous system enclosed within the skull. It includes all the higher nervous centers receiving stimuli from the sense organs. The brain interprets and correlates these stimuli to create your thoughts, feelings, and movements. It is the organ of the mind.

The Mind: The element, or complex of elements, that allows a person to perceive, think, feel, reason, and will. The mind is the manifestation of consciousness.

Free Will: The ability to choose and decide, the freedom of humans to make choices that are not determined by prior causes or by divine intervention. Free will represents the victory of human reason over animal instincts, "I do this of my own free will."

Most people believe we need all three — the brain, the mind, and free will — to describe the way we see ourselves. Our brain keeps our body functioning, our mind is where thinking and feeling occur, and free will allows us to integrate thoughts with feelings to make choices and decisions and to act on those decisions. Most people view the interactions of the mind and the brain as merely unfathomable, whereas the question of how free will leads to action is an abyss that logic and reason cannot begin to bridge.

1

For ages, people have also been fascinated by drugs that alter the mind, and ultimately impair free will. How do drugs do that? How do people become different under the influence of drugs? What makes these differences come about? Many people seem to like the changes produced by drugs such as alcohol, marijuana, and cocaine. Once they try these drugs, many people keep using them, in spite of problems that often develop with increasing use. Moreover, some of those who keep using drugs over an extended period of time become addicted. How does *that* happen? And once people are addicted, why can't they quit?

Enter Neuroscience

Only within the past couple of decades has a relatively new field of inquiry, *neuroscience*, emerged to develop the tools that can begin to answer these questions. Neuroscientists have combined the methods of psychology, biology, chemistry, computer science, and anything else they could get their hands on to explore the workings of the brain. They have brought about an explosion of knowledge. As a result, our ability to think about drugs, the mind-brain interaction, and free will is changing.

A striking irony has emerged from this new knowledge: As drug use progresses to abuse and addiction, drugs rob people of the very free will they exercised when they first decided to use drugs. How this happens is something that neuroscientists are now beginning to understand. But because the language of science is highly technical, few other people have had access to this knowledge. Although neuroscientists don't yet have all the details, they do understand some of the basic mechanisms of addiction. This book explains what neuroscientists know about drugs and the brain in language everyone can understand.

The Brain Reward System

A key to understanding how drugs work lies in a group of interconnected brain parts, which reward you when you do the things you need to do to survive. This brain circuit produces "reward" by making you feel pleasure when you do these things. You eat and drink so that you can survive. Eating and drinking make you feel good. You have sex so that your species can survive. Sex, too, makes you feel good. The feelings of pleasure that result when you accomplish survival-oriented tasks are caused by the release of specific neurochemicals within this circuit, called the *brain reward system.*

Your brain uses these feelings to teach you to repeat these behaviors. Assuming other factors don't intervene, this primitive, yet ingenious, form of learning helps to ensure that both you and the human race actually do survive.

The cells in your brain communicate by releasing neurochemicals. Neurochemicals are your brain's *real* messengers. They help to bring in information from the outside world about loving glances, fierce stares, the state of the stock market, or the score of your child's soccer game in the first half. Your brain's messengers are crucial if you are to process all the information, make decisions about what to do with it, and send commands to your muscles to respond. You may respond with a kiss, stare back, celebrate (or deplore) NASDAQ, or cheer your soccer player on to victory.

DRUGS ARE FALSE MESSENGERS

Drugs of abuse are false messengers. In some parts of the brain, drugs actually send false messages. In other parts, they weaken or intensify real ones. But no matter how they disrupt communications in other parts of the brain, nearly all drugs of abuse mimic the actions of the neurochemicals that make people feel pleasure when their brain reward systems are activated. Drug users describe the intensified pleasure produced by drugs as being "high." In fact, drugs turn on the reward system with a potency that natural rewards can rarely match. Because of this, drugs actually teach people to use more drugs.

Drugs have other effects. They produce *tolerance*, meaning that after a period of use, people need more of the drug to experience the same pleasure they felt at first use. So, at the same time that drugs are using powerful brain circuits to teach people to repeat drug-taking behaviors, the brain and body are responding by adapting to the presence of drugs and causing drug users to take increasingly larger doses to get high. In other words, as drug tolerance develops, many users escalate their drug doses, setting the stage for the development of physical dependence and later, if abuse continues, addiction. As drug use progresses, the brain, our master teacher for survival, must instruct according to the dictates of the intruder that has taken over its classroom. The more that drugs dominate this classroom, the closer people move toward addiction. And the closer people move toward addiction, the more free will they forfeit.

The brain uses a second form of instruction, which not only helps to move people toward drug addiction, but keeps them there when they attempt to stop using drugs. Anyone who has tried to cut down or quit smoking cigarettes will recognize this teaching method. Drug users gradually learn to associate neutral behaviors such as talking on the phone or driving a car with drug-taking, like smoking a cigarette. Pretty soon, like the sound of the bell that made Pavlov's dogs salivate, these neutral behaviors *by themselves* induce craving for the drug.

Neuroscientists' quest to learn how drugs assume control of the brain, of behavior, and, indeed, of free will, is providing us with a new window into the brain. Through that window they are seeing how our brain acts as the seat of our thoughts and feelings and how it controls our behavior. Drugs change behavior by changing the way the brain works. Our emerging understanding of how drugs do this has enormous implications. It forces us to reexamine how we view drug addiction, to reconsider how we see ourselves, and even more profoundly to re-conceptualize the brain, the mind, and free will.

CHAPTER 2

CURRENT DRUG USERS — FUTURE ADDICTS?

*T*o explain how drugs change the brain and produce addiction, we will follow the histories of several people whose involvement with drugs ranges from initial use to full-fledged addiction. Some are addicted when we meet them. Others may not become addicted. Most will. Let's meet them now. (Unlike the people whose letters appear in the preface, the drug users we are about to introduce are not real people, but composites we have created. They are based on the interactions we have had with thousands of drug users and their families.)

Neil and Michelle

Neil didn't smoke that much, but when he wanted to get high, he could depend on Justin to sell him good stuff. Thanks to Justin, Neil was able to show up at Michelle's door tonight with a couple of loosely rolled joints hidden in his shirt pocket. He said his "hellos" to Michelle's parents and then took Michelle outside to lie in the rope hammock facing the trees in the backyard. Neil lit a joint and passed it back and forth to Michelle in cupped hands to hide what they were doing from Michelle's parents.

After finishing the joint, Neil and Michelle hopped into the car and headed for the dead-end road near the bay. Lots of young people gathered there when their parents were at home. Neil and Michelle sat entwined in the front seat of the car. They kissed for what seemed like forever, and Neil could not believe how good it felt. He had kissed Michelle before, but this was different. Must be the pot, he told himself.

Neil was really getting into it when suddenly a harsh, bright light flashed at them through the fogged-up back window. Damn! The police! They both panicked. They were stoned, and Neil still had that other joint in his pocket. But, as the police car spotlight scanned the area, Neil realized that all the cops wanted was for the kids in the cars to leave. He cautiously backed out past the police officers and tried to calm Michelle, joking about being busted on lover's lane.

But Neil's humor masked his own alarm. As he turned onto the winding road that led back to town, Neil could feel his heart pounding. The pot-magnified fear created by the cops' intrusion had released a giant surge of adrenaline, which was interacting with the pot he had smoked. His body was on emergency alert, but he couldn't concentrate. He had trouble judging the distance and speed of oncoming cars. He couldn't maintain a steady speed. He kept going too fast or too slow. When Neil tried to look into the rear-view mirror, he drifted toward the other side of the road into the path of an approaching car. Michelle's scream made him pull back in time to avoid a crash, but he couldn't straighten out the car, and he ran it off the road into the sand. The car stopped abruptly and stalled. Neil started the engine, but couldn't move because the tires were in deep in the sand. They were stuck.

"Now, what do we do," whimpered Michelle.

"Great," Neil thought to himself, as he considered how he'd hoped this night would turn out. "Just great."

Chris

Chris was elated. Just a couple of hours earlier he'd been doomed. The term paper for his economics class was due tomorrow, and he hadn't even started it. No way could he finish it by morning; he was too tired to stay up. But now, his problems were over. He had copped some coke from a classmate, and he knew he could stay up all night and write. He tested his new stash. It was good stuff. His nose went numb almost immediately, and he could taste the bitterness as those first few snorts ran down the back of his throat. In seconds, he felt that familiar light go on inside his head. He was more alert, more confident, smarter. He would get an A!

Chris started writing. It was hard to keep a good train of thought, but his confidence grew with each hit he took. One inspired thought after another flew around in his brain. It was all he could do to get his thoughts down on paper. It was all he could do to sit still. This was going to be excellent. His tongue ran in and out of his dry mouth. He was going to send this to *The New York Times!* He jumped up and down out of his seat as he typed. The White House would want to read it! He paced around the room. He would be a Rhodes scholar!

As the night wore on, Chris's elation wore off. He got more and more anxious. A creaking noise on the back stairs really annoyed him. Then his coke ran out. It didn't matter. He had finished the paper. He ran spell-check and grammar-check, and printed the sucker. . . .

Chris was stunned when he got his paper back the following week. A "D"! At first he didn't believe it. That idiot teacher had always had it in for him! Then he read his paper. He couldn't believe what he read. The beginning was okay. It presented a couple of good arguments, but the paper was disorganized, repetitive, and — he just had to admit it — some parts were downright stupid. The sentences were strung together in a random order. He hadn't come back to the good arguments he had started. Now what would he do? Chris wanted a beer in the worst way, but he had spent his next 2 weeks' cash on the coke.

WHAT DO NEIL, MICHELLE AND CHRIS HAVE IN COMMON?

Neil, Michelle, Chris, and the other drug users you will meet in this book are using drugs that affect the most important organ of their bodies — their brain. Your brain is your body's control center, and it constantly helps you adapt to your worlds — the one around you and the one inside you. Your brain has many parts, and each part performs a different function. Throughout this book, we will focus on a key part, the *brain reward system*, on which drugs of abuse exert their most important effects. This part of your brain releases neurochemicals that make you feel good when you do the things you need to do to survive (eating, drinking, having sex, and so on). The activation of your brain's reward system ensures that you will do all of these things by making you feel pleasure when you do them. *Your brain's reward system teaches you to do the things you must do to survive.*

Drugs are false messengers that produce pleasure by either weakening or intensifying the neurochemicals that normally activate the brain reward system or by replacing those neurochemicals and activating the system directly. By falsifying the brain's natural messages of reward and infusing drug users with artificially powerful feelings of pleasure, drugs reinforce drug-taking. But, unlike the brain's neurochemicals, which can help to teach us to prefer certain behaviors but don't addict us, drugs can cause addiction. They do this gradually by changing the way the brain works. The changes occur because the brain slowly adapts to the presence of drugs and their effects. As this adaptation takes place, drug users become aware that they must keep taking drugs, initially to feel good, later to stop feeling bad, and, eventually, to function "normally."

People do not start using drugs to become addicted. They use them because drugs first produce pleasure, or relaxing, short-term effects. The cocaine that Chris used is a good example. Although cocaine produces feelings of pleasure, it also alters perception, judgment, and thinking and has powerful effects on mood. In this instance, cocaine gave Chris a false sense of confidence. He was certain he was writing an "A" paper, but when he read it later, without cocaine in his system to cloud his judgment, he saw how wrong he was.

Chris is lucky. He has used cocaine only a few times, and so far the worst thing that has happened to him is getting a "D." Both cocaine and crack (a different chemical form of cocaine, which can be smoked) can have fatal consequences, sometimes after just a few uses. Moreover, because of the intense craving that both forms of the drug create in the brain, they can cause very powerful addictions.

Some drugs not only make people feel good, they also make them feel less bad, that is, they produce mild euphoria and relieve negative feelings. These drugs act on the brain to reduce anxiety, allay stress or relieve other negative feelings. If you were to ask Neil and Michelle why they smoke, they would say for the thrill of it. But they also took the drug to overcome the anxiety that emerging sexual feelings often produce in adolescents (and sometimes in experienced adults, even though these feelings are not new).

Marijuana also impairs concentration and distorts one's sense of time, factors that contributed to Neil's disabling his parents' car. Neil and Michelle are relatively new to drug use. Whether they stop using altogether or progress to more intensive use will depend on a number of factors, which we will explore as we examine how drugs teach the brain to take more drugs.

Allison

Allison didn't know how she could stand flying home the next day. Her visit to her grandparents' home had been the best summer of her life, the summer she fell in love. God, she was going to miss Barry.

"Want another beer?" Barry asked, as he lay down his joint and leaned over the blanket to reach for the cooler.

"Yes," Allison said.

She knew she shouldn't. She'd drunk too much already. Five, maybe six cans. A six-pack! But why not? After all, this was their last night together, the last time they'd spend at the beach, drinking beer, trying to keep warm beside a dwindling fire.

Some of Allison's new friends had sex for the first time this summer. But not Allison and Barry, even though they came close once or twice. She had made the decision to postpone having sex, at least until she was through high school. First, it was her choice, which she decided after a great deal of thought and soul-searching. Second, it was way too embarrassing to get protection. She just couldn't make herself go to the school clinic and ask for the pill. Or, worse, carry condoms around in her purse. What if her mother saw them — oh, God — or her father!

Her decision not to have sex had been easy . . . until she met Barry. She had held him (and herself) off, but now, tonight, their last night together, she wasn't so sure she still wanted to. She was pretty drunk, and the beer was beginning to erode her resolve not to give in to the feelings that Barry's closeness stirred up in her.

Barry pulled her over, kissed her, and began moving his hands toward places she hadn't let him touch before. She couldn't tell if it was his touch or the beer that made her so dizzy, but explosive feelings overwhelmed her. Somewhere in her head a voice warned that they were not protected, but another voice said "just once won't matter," and she heard her own voice saying, as if it had a mind of its own, "Don't stop, Barry. Oh, God! Don't stop."

ALCOHOL'S EFFECT ON ALLISON'S JUDGMENT

The drug Allison is using — alcohol — also acts on the brain's reward system, but it affects many other parts of the brain as well. Because of these effects, alcohol can cause noticeable changes in the drinker's behavior after just a few drinks. The six-pack that Allison drank impaired the function of her cerebral cortex, the part of her brain that allows her to exercise good judgment. She did something she would not have allowed herself to do if she had been sober, and she violated a code of behavior she had established for herself and adhered to until that night on the beach. In just a single episode of drinking, alcohol can produce acute changes in the brain and alter behavior in ways that can have serious consequences. The drug can also cause serious long-term changes, including addiction, permanent memory loss, liver disease, and many kinds of cancer. These long-term changes occur when people gradually increase the amount of alcohol they drink as they develop tolerance to it, become physically dependent on it, and eventually become addicted to it.

Not everyone who drinks becomes addicted. In clear contrast to the alcoholic, many people can drink moderate amounts of alcohol throughout their lives and experience few if any problems.

Henry

Henry felt lucky. He had scored some of the purest heroin that Jake, his dealer, ever had. Just smelling the glistening white powder started to make him feel high and brought on a tremendous desire to shoot up right there in the middle of the street. But somehow he made it home.

Henry's wife hated his "habit," and he rarely shot up when she was home. But tonight he couldn't wait. He walked past her protests and sneering insults, closed the bathroom door, and got out his "works." All the veins in his arm were ruined, but in a few seconds he found a good vein in his foot and injected himself.

Henry knew something was wrong even as the rush hit him, and he tried to get up and walk out of the bathroom. He fell against the door and knocked it open, falling to the floor in front of his startled wife. His breathing was slow and shallow. Deeply frightened, she called 911.

By the time they got Henry to the emergency room, the paramedics had already started to perform CPR [cardiopulmonary resuscitation]. Nonetheless, Henry's lips had turned blue. The resident could not detect any breathing, although Henry's heart was still beating strongly. The ambulance driver had brought along Henry's works, and Henry's wife confirmed he was an addict. The resident made a quick decision and administered naloxone.

In a few moments, Henry started to breathe again and the blueness disappeared from his lips. Soon he was sitting up, asking to go home.

HENRY'S NEAR-LETHAL ALLIANCE WITH HEROIN

As we see with Henry, who overdosed on heroin and nearly died, drugs like heroin, cocaine, and alcohol can have lethal, short-term effects. By mimicking a natural neurochemical in Henry's brain, heroin blocked the messages his brain uses to regulate his breathing. If it hadn't been for the paramedics, he would have stopped breathing. A heroin-blocking drug, the antagonist naloxone, which scientists originally developed in an effort to produce drugs that would relieve pain but not produce addiction, saved Henry's life. An antagonist is a drug that blocks the effects of another drug. In Chapter 5, we will explain how antagonists work in the brain to block the effects of drugs and how they can help addicts recover.

Henry is addicted to one of the most severely addictive classes of drugs known to man, opiates. After opiate addiction becomes firmly established, withdrawal symptoms are so pronounced and craving for the drug is so powerful that few addicts succeed at recovering without experiencing several relapses before they reach their final goal, if they *can* reach it. Because of its power, scientists have studied opiate addiction longer than they have studied addiction to other drugs. As a result, much is known about opiate addiction.

Another reason why opiates have fascinated scientists is that throughout history opiates have been the most potent pain relievers known to man. Scientists have naturally focused on these drugs in a dual effort to help addicts overcome an addiction that is so powerful it often seems intractable, and, at the same time, to find more effective analgesics.

As a result, we know more about how opiates act on the brain than we know about most other drugs of abuse. In fact, the breakthrough that accelerated research on drugs and the brain occurred during a search to try to understand how opiates work on the brain. This search led to the discovery of a unique set of molecules in the brain, called *opiate receptors*. Opiates attach themselves to these receptors to exert their effects. The discovery of the opiate receptor in the early 1970s ignited an explosion of knowledge that has continued ever since.

Sybil

"Dr. Schwartz for you on line 3," said Sybil's assistant.

"I'll be right with him," Sybil replied, checking her watch.

Two hours were left before Sybil had to leave for her flight. Time enough to finish her speech. Her bags were packed and in the car. She reached for her cigarettes. Damn! Only half a pack left. She'd have to buy more on the way to the airport. Both her staff and her husband, Daniel, had been "on" her about her smoking. They were right. Everyone she knew had quit. She was the only one left who still smoked. One day soon, she told herself, pushing the button for line 3.

"Good morning, Jerry. What's up?"

"I've got the x-ray from your check-up, Sybil. I want you to come in. Right away," he added, before she could contradict him.

"Is everything all right?" she asked, a sudden coldness gripping her. Too alarmed to hear his answer, she cut it off. "Never mind. I'll be right there," she added, hanging up.

Dear God, what could be wrong? At 48, she was at the height of her career, with a husband she loved, kids safely launched, one in college, one at work. She grabbed her laptop — she could finish her speech on the plane — and hurried down to her car....

"There's a spot on your left lung," Jerry said, showing her the small dark gray shadow on the x-ray.

She could hardly get the words out. "What does that mean?"

"It means you may have lung cancer. I've scheduled a CT [computed tomography] scan in the morning and an appointment afterward with Dr. McCloud. Best lung man in the city. St. Joseph's Cancer Center. Tomorrow, 8:00 AM," he said. Each word hit her like a slap.

"What? But I'm supposed to go to Dallas to give a speech. I ca –"

He stopped her. "It can wait. You can't," he said, putting his hand on her shoulder to still her shaking, as the coldness inside took over completely.

SYBIL'S LUNGS MAY PAY A PRICE

Sybil uses the drug that far more Americans are addicted to than any other — nicotine. Her addiction is not unusual: the surgeon general states that 9 out of 10 people who smoke are addicted. Sybil has been addicted to nicotine most of her life, and she certainly needs it to function normally. Each time she inhales, the cigarette smoke she draws in delivers nicotine to her brain. Unfortunately, it also delivers other poisonous chemicals to all parts of her body. Over time, these chemicals help to create heart disease, lung disease, and an astounding array of cancers throughout the body. Because Sybil has satisfied her constant need for nicotine for a very long time, she has exposed the cells in her lungs to chemicals that changed them. Now these changes may have led to lung cancer.

* * *

As you learn about what is happening to Neil, Michelle, Chris, Allison, Barry, Henry, and Sybil, it becomes apparent that to understand how drugs have changed their brains and produced the behaviors they are engaged in, you must first understand how the brain itself works.

As we explain throughout this book how the brain communicates and how drugs change the brain's communications system, we hope that you will gain a profound appreciation for this miraculous organ that makes you everything you are. Your brain enables you to think, to plan, to imagine, to dream, to create, to form new relationships and sustain old ones, to be everything you want to be, and to do everything you want to do in life. It also manages your body functions and is the source of all of your thoughts and feelings. It enables you to make choices and decisions and to plan and carry out your movements so you can act on those decisions. In short, it governs how you behave.

While drugs of abuse may affect any organ in the body — and different drugs can affect different organs — *the one thing that all drugs have in common is that they affect the brain.* When we consider what risks drugs present, the most serious ones all result from a primary effect on the brain. If a person dies from a drug overdose, it is probably due to the drug's effect on the brain. If a person has a car crash or other accident while high on drugs, these calamities are the results of drugs' effects on the brain. If a person becomes addicted, it is because of the way the drug changes the brain. If long-term damage to other organs occurs, it is because the addict's addicted brain would not let him or her stop using drugs.

As we follow the drug users we have introduced here, we will explain how the drugs they are taking change the way their brains process information and, as a result, change and gradually gain control of their behavior. Because drugs change the brain directly, they affect everyone who uses them — man or woman, black or white, rich or poor, young or old. Ultimately, if drugs are used long enough, the changes they produce can rob people of their free will.

CHAPTER 3

HOW THE BRAIN IS ORGANIZED

*T*o understand how drugs change the brain, it helps to first understand how the brain works under normal circumstances. So let's focus on you and your brain.

You live in a complex, ever-changing world. Your brain must keep you alive and functioning effectively in this world. For instance, your brain tells your heart to beat at a speed that is healthy for you and tells your lungs to supply you with oxygen to keep you going. It also makes it possible for you to see, hear, and feel what is taking place around you and then uses this information to guide your actions. What's more, all these conditions change moment by moment. Your brain must constantly monitor them so that you can constantly adapt to them.

For instance, getting up from a chair changes your heart rate. Actually, every movement of your body and head changes the information coming in through your eyes, ears, and skin. Your brain needs to decipher this barrage of information so that it can help you decide which of the things going on around you is most important and then deal with those things effectively.

To live in society, you must relate to other people, plan for the future, and, to avoid Neil's fate (see Chapter 2), stay on the right side of the road when you drive. A normal brain easily does *all* these things at once, although most of us are rarely aware of concentrating on more than one.

DRIVING — A COMPLEX TASK TO DESCRIBE, A SIMPLE TASK FOR THE BRAIN

Let's examine what goes on in your brain when you perform a familiar activity such as driving a car. Driving involves many parts of your brain. For example, every time you get behind the wheel, you engage in a massively complicated job that requires you to deal with a huge amount of sensory information, to identify what is important in that information, to use the critical elements to make split-second decisions, probably to juggle several other cognitive tasks, and to keep your emotions in check (e.g., when your kids in the back seat are fighting about who gets to hold the dog).

Despite the complexity of this undertaking, you usually do it without paying a whole lot of attention to what you are doing and no attention at all to what's going on inside your body. To most people, driving seems pretty mundane. But only the most powerful computer could carry out any one of the different tasks you must accomplish to drive, and no computer has yet been designed that could combine all those tasks to drive a car.

LOCALIZATION OF FUNCTION

Your brain can cope with complex tasks because of the way it is organized. The rules that govern its organization evolved along with the brain itself. One of these rules is particularly useful to consider, because it makes it easier to understand how drugs act on the brain. This rule, called *localization of function*, says that specific places in your brain carry out specific functions of your brain. For example, your brain directs the information coming from your eyes to a region toward the back of your head (Fig. 3.1). This part of your brain, appropriately called the *visual cortex*, receives information only from your eyes and does nothing else but analyze what you see. Without the visual cortex, you could perceive nothing of the visual world, even if your eyes were healthy and working properly.

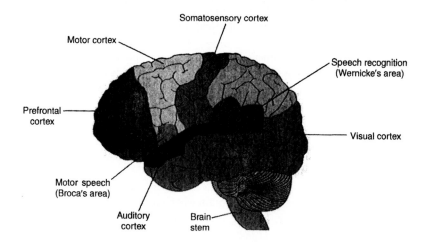

Figure 3.1 Localization of function in the cerebral cortex. Each major region shown here consists of many cortical fields. The visual cortex may contain several dozen by itself.

Another region, farther forward in your brain, called the *somatosensory cortex* (from the Greek *somato* meaning body), processes information coming from your muscles, joints, and skin. It gets information only about what you feel on your body surface and in your muscles and joints. This information, which is very important for driving, is called *kinesthetic information* (from the Greek *kinein* to move). It tells you how your various body parts are oriented in relation to each other so that you can move them accurately. Kinesthetic information can tell you how the angles of your wrist, elbow, and shoulder change as you turn the steering wheel of your car or shift gears. Thus, these two different kinds of information — one about what you see and the other about your skin, muscles, and joints — is directed to two different regions that process and perceive them independently of each other. Another part of your brain, the motor cortex, eventually combines these two kinds of information so you can generate accurate movements just when you need them.

Different functions take place in different parts of your brain because information is directed with great precision. Because your brain sends information to exclusive regions, the kind of information it receives actually determines what that region can do. This is why you see stars, for example, when you get hit in the eye. That blow is hard enough to activate the unique nerve cells in your eye that ordinarily have the job of sending visual information to the brain. These specific nerve cells change light into a special set of coded signals that your brain can understand, but they are sensitive to pressure as well. Every day, these cells transmit billions of bits of information about light, and your brain always interprets information coming in from these cells as if it were light (it almost always is). So, even if pressure turns these cells on, your visual cortex still interprets this information as if it were created by something you saw. As a result, a good smack in the eye, or even rubbing your eye, can make you see "stars" as pressure inappropriately turns on your visual pathway.

So far, we have spoken of the visual cortex and the somatosensory cortex as if they are each single structures. This is not so. Rather, each of the broader areas of the cortex that encompasses the visual and somatosensory cortex are subdivided into smaller parts that do detailed work on sensory information. In fact, neuroscientists who study the visual system now believe that your visual cortex alone contains several dozen subdivisions. Each of these subdivisions is called a *cortical field* and has its own specialty. One cortical field analyzes shape, another analyzes color, and a third analyzes location or motion. In humans, and some other animals, one cortical field is specialized to analyze faces. A man who had an injury to this particular cortical field in his brain could not recognize his own wife from a photograph, but easily recognized her from her voice.

PARALLEL PROCESSING

Even though your brain consists of billions of nerve cells, no single nerve cell works alone. These cells are organized into thousands and thousands of groups. In the cerebral cortex, the groups are called cortical fields. Everywhere else in the brain, they are called *nuclei*. No matter what they are called, each group of nerve cells concentrates on a small aspect of a bigger job. To get the big job done, all the relevant cortical fields or nuclei work simultaneously on their own small parts of that job. Computer scientists call this *parallel processing*. A construction crew might simply call it teamwork.

To understand how your brain divides up work to get it done efficiently, you can compare the way your brain handles information with the way a construction crew builds a house. For instance, a crew consists of many different kinds of craftsmen. Each kind of craftsmen has a special skill and performs just one kind of job. Similarly, although your brain contains many cortical fields and nuclei, each one handles just one kind of information. Also, craftsmen often work in only certain parts of a house. In your brain, the rule of localization of function determines that specific places always found in the same part of the brain carry out specific functions. Finally, much of the crew works on a house at the same time to get the job done faster. Cortical fields and nuclei all work on incoming information at the same time.

Let's continue with the analogy. Plumbers, electricians, and carpenters do their special jobs simultaneously so they can build a house more rapidly. And because the craftsmen in each group are experts at their own craft, the whole job gets done more skillfully. In addition, plumbers work only with pipes, electricians work only with wires, and carpenters work only with wood. The kind of materials each craftsman works with actually helps define his job. In the same way that materials help to determine what each craftsman may do, the kind of information a certain brain region receives helps to determine its job. Just like the crew that builds a house, your brain splits up every job it has to do into smaller tasks and performs all these tasks in parallel — at the same time. Like craftsmen, the specific regions of your brain that handle these tasks specialize in carrying out just one function.

TYPICAL TASKS YOUR BRAIN PERFORMS WHEN YOU DRIVE

Before we actually consider how your brain is organized to carry out its tasks, let's look at some of the things it must do during a task such as driving.

For example, you are driving to the supermarket to buy food for a special dinner you are going to prepare for important guests.

First of all, you can't drive if you are asleep. Therefore, your brain must control all the different parts of your body that keep you awake and alive. At the same time, your brain must prevent your bodily needs from interfering with your driving. It cannot, for example, let you fall asleep at the wheel or let you run out of oxygen because you stopped breathing. So your brain must keep track of these important, but uninteresting, "housekeeping" chores.

While your brain attends to these and other bodily functions, it also helps you perform other tasks important to driving. You have to see and react to oncoming cars, find the right streets, and sometimes react to unexpected emergencies. You have to control the car under changing traffic conditions. When driving conditions are easy, you may do all these things thoughtlessly, but when they become more difficult you have to focus your attention. All this requires processing a great deal of sensory information — mostly visual and kinesthetic — and turning that information into commands to your muscles so they can control the car properly and get you to your destination.

When you reach the supermarket, you must make a left turn in heavy traffic to get into the parking lot. You need to judge the speed and distance of oncoming cars, calculate how long it will take to cross the traffic lanes, and switch your foot from the brake to the gas pedal to get the car moving. If you are driving a car with a manual gear shift, you have to depress the clutch with your left foot as you control the gas with your right, steer with your left hand, and shift gears with your right hand. Four things at once! If you had to think about each one separately, you would have a tough time. Remember how hard it was to do this when you were first learning to drive? You had to concentrate very hard to do all these things, and you probably did not always succeed at getting them just right.

(One of the authors of this book remembers the time he caused one of his father's trucks to stall in front of oncoming traffic when he was trying to turn left at a busy intersection in Brooklyn. He had let out the clutch too quickly, and the truck suddenly lurched forward and died. Because the truck was bigger than any of the oncoming cars, they all stopped until he could get the truck out of their way. He eventually learned how to use the clutch and has no recollection of ever stalling in traffic since then. Like most adults, he now makes all these driving-related movements smoothly, easily, and with little mental effort. His brain does all the work for him.)

Actually, driving to the supermarket doesn't involve only four things at once. While your brain takes care of all the sensory and motor jobs directly related to driving, it does other things as well. You are probably thinking about tonight's dinner. You may be talking on a car phone, turning the car radio to the stations you like rather than the stations set by the kids or your spouse, or trying to figure out a problem you have at work. All this goes on simultaneously while you drive. And you *can* do all these things at once because your brain divides up the work into small jobs and distributes the information needed to perform each one to the pertinent specialized regions.

Finally, your brain has one more job to do for you as you drive to the supermarket. It monitors your thoughts, memories, needs, and wants and creates the emotional and motivational context for all of your activities. Are you nervous about this important dinner? Are you late? Will there be fresh tomatoes in the produce section for your spaghetti sauce, or will you have to change the menu at the last minute? Although these thoughts are not related to driving, your brain nonetheless produces these kinds of thoughts as you think about — and sometimes even when you don't consciously think about — all the different things you have to do. They all influence how you behave. And they can influence how you drive. For instance, if you are in a good mood, have plenty of time, and have ordered the key ingredients (including fresh tomatoes) ahead of time, you can relax and be a courteous, defensive driver. But if you had a fight with your kids (the radio stations they set just make you more angry) and you are late and running out of time to prepare the meal, you are likely to be in an entirely different mood. This mood will affect how you drive. You may be tense, impatient, distracted, and not particularly courteous. You may even be an aggressive, offensive driver, at least for this trip to the supermarket. You're going to make that left turn into the parking lot no matter what. You may charge through a smaller gap in traffic than usual, causing the oncoming cars to slow down quickly.

Our brains create emotional and motivational aspects for our lives that can profoundly affect the way we handle our sensory and motor tasks. The emotional and motivational aspects of our lives are also key components in drug abuse and addiction.

THE BRAIN STEM

In the previous text, we have considered the many different tasks you need to accomplish while driving to the supermarket. Let's see how your brain actually handles these tasks. By doing this you can learn much about the way your brain is organized and thus lay the foundation for understanding how drugs work.

Let's start with your survival needs. A part of your brain called the *brain stem* has primary responsibility for these needs (Fig. 3.2). Your brain stem is a relatively primitive brain structure that starts where your spinal cord enters your head. All vertebrate animals (i.e., animals with backbones, such as fish, snakes, frogs, birds, rats, cats, and humans) have a brain stem. Their brain stems look pretty much like ours and do pretty much the same things.

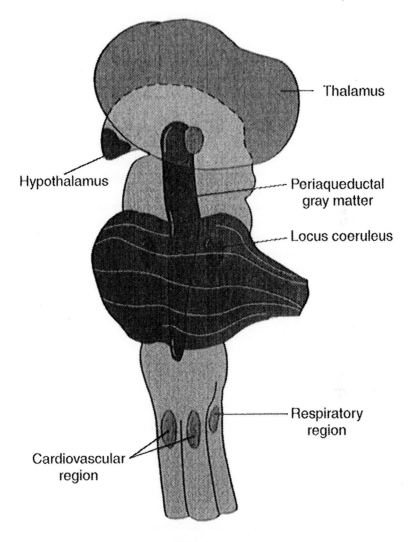

Figure 3.2 The brain stem and thalamus with the cerebral cortex removed. At the top of the figure, the brain stem is actually behind the thalamus, which has been made transparent for this illustration. Only a very few of the nuclei located in the brain stem are shown.

Different parts of your brain stem monitor and control each important function your body must perform to keep you alive and working well. One part keeps track of your heart and blood vessels so that it can keep your heart beating at the proper rate and keep your blood pressure constant. Another nearby part of your brain stem gets information about the amount of oxygen and carbon dioxide in your blood so that it can determine how quickly and deeply your lungs must breathe. This respiratory center in your brain stem is ultimately responsible for telling the muscles that move your chest and diaphragm (the muscles of respiration) how hard to work. Other parts of your brain stem keep you awake when you need to be awake and put you to sleep when it's time to rest.

Still other parts in a structure called the *hypothalamus*, which occupies part of the brain that sits on top of the brain stem, make you feel hungry or thirsty when it's time to eat or drink, so that you'll get the food and liquids you need to survive. Nearby parts, also in the hypothalamus, help to control the sexual urges that enable our species to survive.

Because all the needs just mentioned are crucial to your survival, they have powerful effects on your behavior. For example, it eventually becomes impossible to resist the strong urge to breathe after you have held your breath for a while. Similarly, when you are very hungry, your thoughts may be dominated by food, and you may find yourself in front of the refrigerator without consciously deciding to go there. If you go to the supermarket when you are hungry, you will probably buy more snack food than you intended. That silent urge to grab a box of cookies is your hungry brain at work! But when you drive, your brain stem regulates most of these housekeeping functions so you can pay attention to the road.

THE CEREBRAL CORTEX

Although your brain takes care of these housekeeping functions so you don't have to think about them, you still have to perform many complex sensory and motor tasks when you drive. The sensory-processing needs alone are daunting. You need to see the road, other cars, and traffic signals; feel the gas and brake pedals so you can press on them precisely; flip the turn signal lever without looking; hear emergency vehicles; and attend to dozens of other sights, sounds, and movements. All these sensory tasks require you to process lots of information from your eyes, skin, muscles, joints, ears, and other sensory organs.

Your cerebral cortex (or more simply, your cortex) carries out nearly all the complex sensory and perceptual tasks involved in driving. Your cortex consists of the large, deeply folded outer layers of your brain that make your head so large. Because your brain has perfected ways to split up jobs, different parts of your cortex carry out different aspects of the overall job of driving. As depicted in Figure 3.1, the visual cortex contains the parts devoted to vision; the auditory cortex contains the parts devoted to hearing; the somatosensory cortex contains the parts that process information about touch and kinesthesia; the motor cortex uses much of the sensory information to create commands that make your muscles move; and the prefrontal cortex works with all of these regions and other areas to allow you to plan and to carry out other cognitive functions. We call these cognitive functions, such as analyzing problems and planning their solutions, *thinking*.

For driving, the visual, tactile, and motor areas of your brain make it possible for you to see the road, feel the steering wheel and floor pedals, and then tell your body how to react to what you are seeing and feeling so you can keep the car on the road and get to the supermarket.

We earlier illustrated the rule of localization of function when we described the visual cortex and the somatosensory cortex. This rule holds true throughout the rest of your brain as well. Each part of your brain, whether it is a cortical field devoted to color vision or a nucleus in the brain stem that deals with a more basic function, such as heart rate, has just *one* main job. If that particular part of your brain is damaged, that particular function, and no other, is impaired or completely lost.

Basal Ganglia and Nucleus Accumbens

Many regions of your brain stem and other brain structures that lie below your cerebral cortex work together with the cortex on certain kinds of jobs. Some of these structures receive information from sensory organs, such as your eyes and ears, and relay this information to your cortex. The *thalamus* is such a structure (see Fig. 3.2). It lies between your brain stem and your cortex, and it acts as a gateway to your cortex for almost all sensory inputs. Other structures, such as the basal ganglia (Fig. 3.3), also work along with your cortex. Your basal ganglia are a rather large, complex set of structures that are involved in generating movements, some cognitive functions, and emotional and motivational activities. Your basal ganglia and your cortex share information and quickly shuttle it back and forth to refine your movements, thoughts, and feelings. This sounds like an exception to the rule of localization of function, but it isn't.

Your basal ganglia are as precisely organized as your cortex, and specific functions are localized to specific parts, just as they are to your cortex. One

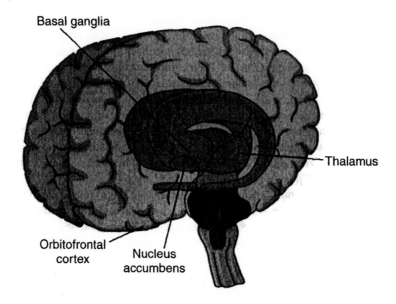

Figure 3.3 The basal ganglia and thalamus as seen with the near hemisphere of the cerebral cortex removed.

part of your basal ganglia, the *nucleus accumbens*, deals with motivation. It is the key brain site where almost all drugs of abuse act (Fig. 3.4). When people take addictive drugs such as cocaine, these drugs act on the nucleus accumbens to produce their pleasurable effects.

Brain Circuits

You have learned that you can point to a specific portion of your brain and say what it does. But because the world is complex, so is your brain. It actually takes more than one cortical field or nucleus to handle a job. In fact, no single part of your brain handles a job by itself. Most functions depend on circuits that link together teams of cortical fields and nuclei. Some circuits are large; others are smaller and less complex. As the job becomes more complicated, escalating in complexity from reading (seeing), to watching TV (seeing and hearing) to driving (seeing, hearing, feeling, and moving), the circuits themselves link up with other circuits to form even larger constellations of brain regions that all work together to complete complex jobs.

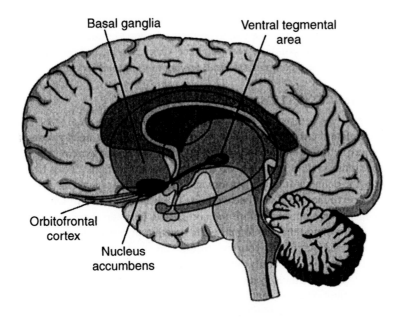

Figure 3.4 The brain reward system. Key components are the ventral tegmental area (a part of the brain stem), the nucleus accumbens (a part of the basal ganglia), and the orbitofrontal cortex (a part of the cerebral cortex). The thick line extending from the ventral tegmental area is a bundle of axons. Some of these axons terminate in the nucleus accumbens; others continue forward in the brain to terminate in the orbitofrontal cortex. These are indicated by single lines.

One way to think about this process is to compare the brain with a large architectural firm that has offices in many cities. A group of architects in the Atlanta office handles the design of an office building. This group can be considered equivalent to a cortical field that processes an aspect of vision, such as color. As the job becomes more complex, like designing a large office park, several groups of architects in the Atlanta office may team up to design various office buildings whose styles will complement each other. This team is like a brain circuit. For the largest jobs, such as designing a new town to support the office park, teams from Atlanta, Winston-Salem, New York, and Chicago may get together to pool their expertise. The Winston-Salem group plans a transit system. The New York group sets up the utilities. The Chicago group lays out the town itself. To get this still larger job done, these teams gather in meeting rooms in all their cities and link themselves together via conference call or video conference. Just as these teams link up by conference call to do their job, brain circuits communicate with each other to combine knowledge of vision, touch, and movement so you can complete a complex task like driving to the supermarket.

THE LIMBIC SYSTEM

To complete the overview of how your brain works, you need to consider one more set of structures in your brain called the *limbic system* (Fig. 3.5). Your limbic system creates your feelings and motivations. Your feelings supply the contexts for your sensory and motor activities and can alter how you perceive the world and behave in it. The limbic system, which evolved after the brain stem but before the cortex, sits between them. Its location provides an important clue to part of its function: it connects your survival-oriented brain stem with your cognitively oriented cortex. While each of these two regions generally lets the other know what functions they are working on, your limbic system uses information from both to generate your feelings, emotions, and motivations, things that reflect your internal states, your memories, and the world around you.

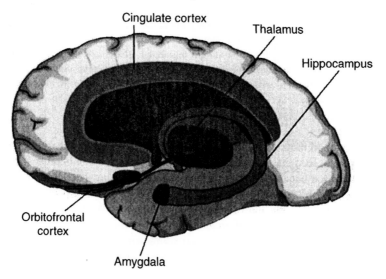

Figure 3.5 The major structures of the limbic system as shown on the medial (inside) surface of the brain. The different components of the limbic system are intimately connected with each other and with many parts of the cerebral cortex and brain stem.

Feelings influence your perceptions in important ways. A sunset over the highway may be so beautiful that it calms you if you take time to appreciate it. But it may just as easily make you angry if you are late and the setting sun is in your eyes. A child's playful antics may make you angry if you are tired, hungry, and busy preparing dinner. After you have eaten, are relaxed, and can pay attention, the same behavior may delight you.

Your limbic system not only changes your feelings as circumstances change, it also changes your motivation. And because it can do this, your limbic system has powerful effects on your behavior. It's the part of your brain that tells the rest of your brain what to do. For example, let's say you ate lunch at 11:30 AM. Toward 4:00 or 4:30 PM, your hypothalamus notes a diminishing level of nutrients in your blood and sends this information to your cortex and limbic system. You recognize this information as the feeling of hunger. As time passes, the signals from your hypothalamus get stronger, and your limbic system sends increasingly more forceful commands to your cortex, which is aware of your hunger but won't do anything about it until ordered to do so. Eventually, in response to commands from your limbic system, your cortex swings into action and you go to the refrigerator (food-seeking behavior).

Thus, your limbic system has taken information about an internal state (hunger) from your brain stem and used its connections with your cortex to get you to satisfy your need for nutrients.

Just as your brain stem monitors your internal states, especially those having to do with your survival, your limbic system generates the commands that get you moving to meet the survival needs your brain stem identifies. Your cortex must figure out exactly what you have to do to get those needs met and then get your body to do those things. For most humans today, meeting your hunger needs is as simple as going to the refrigerator. For your ancestors, it was much more complex. They had to hunt. Their limbic systems supplied the commands that initiated the hunt and motivated them to spend long periods of time outside in the cold, wet woods to get the required results — dinner!

Not all of your decisions to act arise from some basic survival need. You make many choices and decisions about activities that appear to have nothing to do with survival. You read books, watch TV, think, take walks on the beach, go to museums, and attend basketball games. Instead, these activities seem to meet cognitive needs that your cortex is more likely to generate than your brain stem. But even if your cortex makes a decision about some activity it would like to pursue, your limbic system still has to tell your motor system to get you going. You can "decide" to watch a cartoon rather than a documentary on public television or vice versa, but in either case, your brain must relay this decision to your limbic system, so it can alert your motor system to get you up from the chair to search for the remote control to turn on the TV and select the proper channel.

After your limbic system informs your cortex of the decision to act, your cortex can do all the things that must be done to let you carry out the action. But unless the limbic system relays the commands to act — and the requests for those commands can originate in your cortex as well as in your brain stem — little is likely to happen. This is clearly illustrated in persons with Parkinson's disease, in whom the parts of the brain that help link the limbic system to the motor system are damaged. Persons suffering from Parkinson's disease typically sit rather still and do almost nothing, although it is clear that they are not paralyzed. They simply don't get going. The disease severs the link between the motivation supplied by the limbic system and the action generated by the cortex.

Under normal circumstances, your limbic system accurately tunes your feelings and motivations as both circumstances and your needs change over time. In the supermarket, anxiety quickly gives way to relief as you spot the ripe tomatoes you need for tonight's spaghetti sauce. This rapid shift from one emotion to another is typical of how quickly the limbic system can change your feelings.

Drugs and the Limbic System

Drugs exert their powerful effects by acting *directly* on the limbic system. They disrupt the careful, moment-by-moment modulation of feelings and motivations that underlie normal behavior. Your feelings lose touch with reality, and artificial relief, pleasure, contentment, and relaxation take over. The ability of the limbic system to control behavior makes drug addiction a particularly difficult disorder to deal with. All the drugs that people abuse have one action in common. *Drugs change the way the limbic system works.* Just as your brain stem signaled hunger and induced your limbic system to create commands to get you to seek food, drugs essentially make the limbic system create commands to get people to seek drugs. This is one of the fundamental reasons why people repeatedly use drugs, even when they clearly understand that drugs can hurt them.

THE BRAIN REWARD SYSTEM

Drugs change the way the limbic system works in a specific limbic circuit that generates feelings of pleasure, which scientists call the *brain reward system*. Because we will discuss the brain reward system repeatedly through-out this book, it is useful to learn a little more about it now.

The core of your brain reward system originates in a group of neurons in your midbrain called the *ventral tegmental area* (VTA) (see Figs. 3.4 and 3.5). These neurons contain a neurotransmitter called *dopamine*. VTA neurons connect to a variety of places within your limbic system, including a part of your cortex, but a key target is your nucleus accumbens in the basal ganglia. When something activates the VTA neurons, they release dopamine into your nucleus accumbens. When this happens, you feel pleasure. A variety of natural events, such as eating when you are hungry or drinking when you are thirsty, turns on the brain reward system. But nothing turns it on with as much force as cocaine, heroin, and other addictive drugs. In fact, the one thing all addictive drugs have in common is that they turn on the brain reward system.

Although different drugs act through different mechanisms, all addictive drugs somehow increase the amount of dopamine in the nucleus accumbens, with profound consequences on people's behavior.

CHAPTER 4

HOW THE PARTS OF THE BRAIN COMMUNICATE WITH EACH OTHER

*W*e have considered how different parts of your brain interact to allow you to perceive, feel, make decisions, and act. Now let's look at how the individual nerve cells participate in the function of the brain.

We have seen that your brain has hundreds of different cortical fields and nuclei, and that each one works on different pieces of a bigger job at the same time. When you drive to the supermarket, for example, your brain stem takes care of your basic survival needs, while different cortical fields constantly process information about road conditions and send out commands to your muscles so you can respond to those conditions. At the same time, your limbic system is busy managing your feelings and motivations. But how, precisely, do the different parts of your brain do this? The answer is, through an intricate communications system that is so massive it takes one's breath away.

NEURONS

Your brain contains an estimated 100 billion nerve cells, which link themselves together into the many populations that constitute the cortical fields and nuclei already described. These special nerve cells are unlike any other cells in the body. They are called *neurons*. What makes them unique is that each one has specialized branches that allow it to make connections with other neurons. Some neurons have so many branches that a single one can connect to hundreds, thousands, or sometimes tens of thousands of other neurons. In fact, neuroscientists have calculated that the 20 billion nerve cells in your cortex alone make so many connections it would take millions of years just to count them!

It comes down to this: your thoughts, feelings, motivations, and actions are the sum of the activity of your billions of neurons and their trillions of connections. When the passage of a symphony is so glorious that it moves

31

you to tears, or a last-second victory by your favorite basketball team brings you to your feet erupting in cheers, or you suddenly understand how to solve a problem, hundreds of millions of your neurons are sending messages to each other across a communications system that is so immense — and at the same time so minute, that it almost defies comprehension.

You may be wondering how there is room for such a massive communications system inside your skull. But, in fact, this communications system is your brain. Its 100 billion neurons, along with an even larger number of special supporting cells called *glia*, make up your brain's 4 pounds of dense tissue. To do their work, neurons must convert information coming in from the outside world into a code they can work with. For example, as you approach an intersection and the traffic light changes from green to yellow, you must see and respond to this so you can stop in time to avoid running into the traffic that crosses in front of you as the light turns red. So, the neurons in your eyes convert the incoming visual information into messages they can send to neurons in other parts of your brain. Once this information has been analyzed, neurons in the motor cortex send precise instructions to motor neurons in your spinal cord, which command the muscles in your legs and feet to get you to step on the brakes.

ACTION POTENTIALS

Neurons' messages have two components: electrical and chemical. The electrical component of this two-part message is called an *action potential*. Each action potential is a brief pulse of electrical current. It lasts less than 2 milliseconds (1 millisecond equals 0.001 second) — much less time than it takes your eye to blink. The fastest action potentials can travel almost the length of a football field in just 1 second. Using action potentials, an individual neuron can quickly speed messages to hundreds of thousands of other neurons.

Your brain turns information from your senses into action potentials for the same reason that the telephone company turns your voice into electrical signals — so your message can travel quickly from one place to another. In fact, almost all our means of communicating with each other — phone, fax, E-mail, TV — convert information into various kinds of electrical signals to process that information and transmit it from one place to another rapidly. Your brain does the same thing. Like modern communications systems, which code signals digitally (that is, in discrete bits of data rather than a continuous stream), neurons send digitally coded messages in the form of action potentials (although the form of electricity used is different). When you drive to the store to get groceries, the neurons in your eyes (which grew out from your brain during fetal development and are therefore really a part of your brain) convert images of the road, such as the surrounding traffic, stop signs, and intersections, into action potentials. These action potentials travel along a pathway beginning with neurons that reside in your eyes. There is one relay at a way station in your thalamus, and the action potentials generated there arrive a few milliseconds later at the first stop in your visual cortex at the back of your brain. Neurons in this part of your visual cortex distribute these incoming messages to other neurons in the many other cortical fields that make up the visual cortex. Together, all these neurons produce a coded representation of the road so you can "see" it. Similarly, other neurons take in all the other information important to driving — the sounds of sirens and horns, the feel of the gas pedal and steering wheel — convert that information into action potentials, and send them along other pathways to neurons in the cortical fields that process sounds, touch, and other sensory information.

Just like the neurons in the cortical fields of your visual cortex, neurons in each of the other cortical fields relevant to driving a car evaluate the information and then send action potentials to neurons in other parts of brain, including your motor cortex. Neurons in the motor cortex, in turn, send action potentials to neurons in your spinal cord to tell them how to control your muscles so you can respond and perform all the driving tasks needed to reach the supermarket.

This immense, hyperactive communications system, which is your brain, enables you to monitor what's going on around you rapidly and effectively. It also allows you to constantly adapt to changing circumstances. You do this by changing your behavior.

As you drive home from the supermarket, you make smooth, subtly modulated adjustments in your muscles to change speed and direction. But you would behave in a rather different way, and more rapidly, if someone suddenly cut you off. Then, your movements would be forceful and very swift: you would stomp on the brake pedal and jerk the steering wheel to avoid a collision. Your brain's ability to change your behavior in a split second is absolutely critical to your ability to take care of yourself.

DENDRITES AND AXONS

To form the neural connections with each other that allow them to communicate, neurons have two special structures unlike those in any other cell in the body. These structures, *dendrites* and *axons*, grow out from the neuron's cell body. As shown in Figure 4.1, they sometimes grow from opposite sides, and this configuration emphasizes their different functions. The dendrites are the branches growing up from the top of the neuron, and the axon is the single cable extending down from its bottom. Dendrites receive information coming in from other neurons, and axons send information to different neurons. At its far end, the axon has its own set of branches, called *axon terminals* (see Fig. 4.1). Some axons have only a few terminals; others have thousands. Some neurons with large numbers of axon terminals contact many other neurons. Others send a very powerful message to a relatively small number of cells.

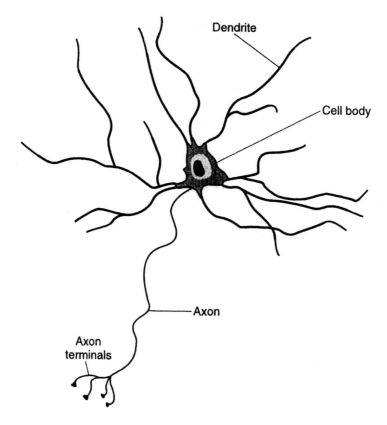

Figure 4.1   Structure of a neuron.

We can compare dendrites with the branches of a tree. Just as tree branches spread out so that all of their leaves can receive sunlight, dendrites spread out to receive signals from the axon terminals of other neurons. In addition, we can compare an axon to a telephone wire. Just as telephone wires have been strung across the country to let you talk to anyone anywhere in the country, axons run throughout the brain and body to send messages to all the nuclei and cortical fields they reach.

A strong message consists of many action potentials being sent very rapidly, in some cases, up to a few hundred per second. If you could hear this message, it would sound like corn briskly popping in the microwave, or a machine gun. A weak message consists of only a few action potentials per second and would sound like a bag of microwave popcorn that has nearly finished popping.

Despite the speed they can travel, action potentials carry only a limited amount of information, signaling only how strongly a neuron has been turned on. Because all neurons use action potentials, the quality of the message — whether you see a light or hear a sound, for example — is determined by the particular neurons that carry it, not by the rate or pattern of the action potentials. (We will make this clearer later.)

A single neuron may receive messages from hundreds, even thousands, of other neurons, and this information arrives almost constantly in ever-changing combinations. At times, a neuron's dendrites can receive what seems like an uninterpretable barrage of messages every second. Despite such apparent chaos, the dendrites and cell bodies can respond to only the most important messages because individual action potentials have little effect on the cell. Only when a number of action potentials are added together, because they arrive at nearly the same time, does the receiving cell respond. We'll explain how this works below.

SYNAPSES

There is a major complication in the brain's communications system that makes it much different from the telephone. If we magnify the brain and zoom in to take a closer look, we see that when axon terminals close in on dendrites to form connections, they don't actually touch each other (Fig. 4.2). A small space, or gap, separates every axon terminal from the dendrite it wants to talk to. These gaps are called *synapses*, and they create an interesting problem for the nervous system. The solution to this problem actually helps each neuron make sense of the barrage of information arriving at its dendrites.

For electricity to flow from one device to another, the wires that carry it must touch each other continuously and securely. If your toaster has ever become partially unplugged and stopped toasting, you know how important

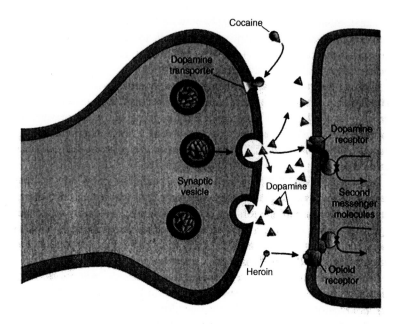

Figure 4.2   An axon terminal (left) releasing the neurotransmitter dopamine into a synapse. This figure also shows that cocaine exerts its effect on the axon terminal whereas opiates, like heroin, work on the dendrite on the post-synaptic side of the synapse.

this is. The toaster's plug must be seated securely in the electrical outlet to make an effective electrical connection. But the connection between neurons is anything but secure. The synaptic gap that separates every axon terminal from every dendrite, creates a space between them that prevents the action potential from getting across, just as the flow of electricity stops when the toaster's plug becomes loose. How does the neuron's information get across this gap when the electrical message can't make the leap? The neuron's solution to this problem is ingenious — neurotransmitters.

NEUROTRANSMITTERS

At every synapse, the neuron changes its electrical signal into a chemical one. When that action potential reaches the neuron's axon terminals, it triggers the release of special chemicals that diffuse across the synapse to carry the message to the next neuron. Scientists call these chemical messengers *neurotransmitters*. Once a neuron receives a neurotransmitter's chemical message, it converts it into an electrical message, which can become an action potential. This electrical-chemical message is repeated every time your neurons communicate with each other. And this dual nature of neuronal communication is what makes us vulnerable to drug abuse. Precisely because the neuron sends a *chemical* to carry its message across the synapse, addictive drugs like alcohol, cocaine, or heroin, which are *chemicals* themselves, can change the way neurons communicate. These false messengers can masquerade as real ones.

Boats and Bikes Model

The following model can help us understand how neurons communicate, so that we can learn precisely how drugs change what neurons say to each other.

Imagine that the brain is an ocean and neurons are islands floating within it (Fig. 4.3). So many islands dot the ocean that there is very little space between them, but some water does separate each island from the others. This water represents the synapses.

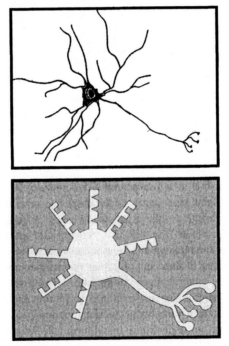

Figure 4.3   Neurons are like islands.

Figure 4.4   An action potential is like a motorcycle.

Imagine further that an action potential is a motorcycle (Fig. 4.4), which speeds down an island road (axon) to the shore (axon terminal) only to discover there is no bridge to take it across the water. Finally, imagine that a neurotransmitter is a boat (Fig. 4.5). The motorcycle (action potential) must somehow transfer the message to a boat (neurotransmitter) to make this part of the journey. Once it does this, the boat (neurotransmitter) can take the message across the water (synapse) to the shore (dendrite) of the next island (next neuron). When the boat arrives there, its journey ends. It must transfer the message back to a motorcycle (action potential) to speed down that island's road (axon) to its far shore (axon terminal). When an action potential reaches the axon terminal of a neuron, how does it transfer the message to a neurotransmitter?

Figure 4.5   A neurotransmitter is like a boat.

VESICLES

Sacks, called *vesicles*, densely packed with neurotransmitters, fill the inside
of each axon terminal. These vesicles manufacture and store the neuron's
neurotransmitters. In some ways, vesicles are like shipyards, which build
and store (dry-dock) boats. When a neuron sends an action potential down its
axon and it reaches the axon terminals, some of the vesicles inside each of
the terminals release their neurotransmitters into the synapse — in the same
way shipyards launch boats into the ocean (Fig. 4.6). These neurotransmitters
travel across the synapse to the next neuron, as boats sail across the sea to the
next island.

Figure 4.6   Vesicles release neurotransmitters into the synapse just like shipyards launch boats into the ocean.

RECEPTORS

Scientists have identified many different kinds of neurotransmitters in the brain. Because each kind has its own distinct chemical structure, each kind has its own distinct chemical shape. The neurotransmitter's shape is important because, like a key, its shape is its signature. But one more piece is needed to complete this puzzle. Every key needs a matching lock to be able to open a door. The part of the neuron that plays the role of the lock is called a *receptor*. Receptors are embedded in the membrane that forms the outside wall of each neuron. One side of the receptor sticks out from the cell membrane on the outside of the cell. The other side of the receptor is inside the cell. This is much like a lock that is mounted in a door. The

Figure 4.7 Neurotransmitters must have the same chemical shape as receptors to be able to bind to receptors, just as boats must be shaped to match docks where they can land. Square boats cannot land in triangular docks, nor can triangular boats land in square docks.

outside portion of the receptor can recognize its designated neurotransmitter.

Dendrites are covered with receptors in much the same way that tree branches are covered with leaves. Incoming neurotransmitters attach themselves to these receptors in a process called *binding*. When a neurotransmitter binds to a receptor that it is shaped to match, it sends a message. If the neurotransmitter can't find a match, nothing happens.

In the boats-and-bikes model, the receptors on dendrites are like docks on piers (Fig. 4.7). The docks (receptors) have different shapes and sizes to accommodate boats (neurotransmitters) of varying shapes and sizes. If a boat can't find a dock it can fit into, it can't land, and it can't deliver its message. But if it does find a fit, the message can get through.

TYPES OF NEUROTRANSMITTERS

Even though the brain has many different kinds of neurotransmitters, the most ubiquitous neurotransmitters have one of just two kinds of messages: excitatory (they help to elicit an action potential) or inhibitory (they prevent firing of an action potential). A neurotransmitter called *glutamate* is the most important *excitatory* neurotransmitter in the brain. Glutamate makes neurons fire action potentials. A neurotransmitter called *GABA* (gamma-aminobutyric acid) is the major *inhibitory* neurotransmitter in the brain. It prevents neurons from firing action potentials. Most of the other neurotransmitters in the brain modulate the actions of glutamate and GABA.

It's as if our ocean contains millions of triangular-shaped (excitatory) and square-shaped (inhibitory) boats (Fig. 4.8). All the triangular-shaped boats have just one message: fire! All the square-shaped boats have just the opposite message: don't fire! This is one way that neurons make sense of all the information that is bombarding them. Other neurotransmitters help neurons make sense of this information by modulating excitatory and inhibitory neurotransmitters (to be described).

Exciting the Neuron

When an excitatory neurotransmitter *binds* to a receptor, it encourages the neuron to fire off an action potential. Although excitatory neurotransmitters tell neurons to fire action potentials, a single excitatory neurotransmitter cannot do the job by itself. A number of excitatory neurotransmitters must bind to a number of the neuron's receptors before the excitatory message is

Figure 4.8 Most neurotransmitters carry just one of two kinds of messages. Excitatory neurotransmitters (triangular boats) tell the neuron to fire an action potential. Inhibitory neurotransmitters (square boats) tell the neuron not to fire an action potential.

strong enough to actually elicit an action potential. The teamwork required to do this is another way the neuron makes sense of all the information it is receiving. A neuron is not even aware of the chatter of individual excitatory neurotransmitters until enough of them arrive together to clarify their message: fire an action potential.

Let's say the captains of each triangular-shaped (excitatory) boat in a fleet are allowed to carry only $1.00 in cash. When the first captain to arrive docks his boat, he tries to hire a motorcycle to speed a message to the other end of the island. But the motorcycle driver says,

"Hey, man. I need 10 bucks to make the trip."

So the captain of the first boat must wait for nine other triangular-shaped boats to land at the docks, so they can pool their money. As soon as $10.00 is collected, they can hire the motorcycle and send the message over land to the shipyards. That is, the excitatory message is now strong enough to elicit an action potential.

The need to have many excitatory neurotransmitters bind to the neuron's receptors at more or less the same time to fire an action potential ensures that only important messages get through. Moreover, a group of neurotransmitters must arrive within a few milliseconds of each other if they are to have enough effect to elicit an action potential.

Inhibiting the Neuron

When an inhibitory neurotransmitter binds to a receptor, it acts to prevent the neuron from firing off an action potential. This inhibition makes it harder for neurons to fire action potentials. Instead of hiring motorcycles when they dock their boats, the captains of square-shaped boats (inhibitory neurotransmitters) hire traffic cops to prevent motorcycles from entering the highway that runs down the island. Like his counterparts on the triangular-shaped boats, the captain of a single square-shaped boat doesn't have enough money to hire a traffic cop on his own. (The message of a single neurotransmitter isn't strong enough to do the job alone.) He also must wait for other square-shaped boats to dock so several captains can pool their money to hire the cop.

DUELING NEUROTRANSMITTERS

Neurons constantly receive excitatory and inhibitory messages from the neurotransmitters that other neurons release. In fact, they are bombarded with them. Excitatory neurotransmitters constantly try to turn neurons on, telling them to fire action potentials. Inhibitory neurotransmitters constantly try to turn them off, telling them not to fire. The piers of the island are a madhouse. Triangular-shaped boats dock at the pier and hire motorcycles as fast as they can (Fig. 4.9). Square-shaped boats dock just as quickly and hire traffic cops

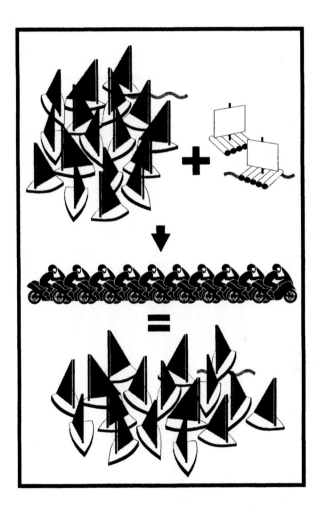

Figure 4.9 When more triangular boats (excitatory neurotransmitters) than square boats (inhibitory transmitters) land in docks (receptors), they can hire lots of motorcycles (fire many action potentials) to release boats at the other end of the island and take the message across the water (synapse) to the next island (neuron).

to stop the motorcycles (Fig. 4.10). Whether an island sends many messages, just a few, or none at all depends on how many square-shaped and triangular-shaped boats arrive on its shores at any particular moment.

In other words, the more *excitatory* neurotransmitters a neuron receives, the more action potentials it fires and the more chemical neurotransmitters those action potentials release to communicate with the next group of neurons. The more *inhibitory* neurotransmitters a neuron receives, the fewer

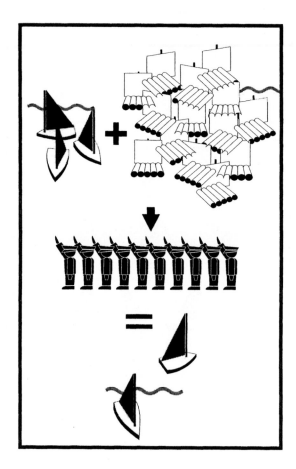

Figure 4.10 When more square boats (inhibitory neurotransmitters) land in docks, they can hire lots of traffic cops to stop the motorcycles. The motor-cycles can't get through, so few boats are released to take the message to the next island.

action potentials it fires and the less it has to say to its neighbors. Our brains need both kinds of neurotransmitters to help us do all the things we do.

SECOND MESSENGERS

The binding of the neurotransmitter to its receptor is the key event in synaptic transmission. Some neurotransmitters, including both glutamate and GABA, bind to receptors that are linked directly to the mechanisms that initiate action potentials. The only thing that has to happen is for a neurotransmitter to bind to its receptor. This is like a key that fits into a lock in a doorknob that unlocks and opens a door with one turn of the wrist. But for most other kinds of receptors, a second step must occur before the binding of a neurotransmitter to a receptor can exert any influence over the generation of an action potential. This is like a door with a dead bolt that has to be unlocked before the doorknob can be turned to open the door.

When these kinds of neurotransmitters bind to receptors, the second step they must take is to create an additional communications link inside the cell to affect the mechanisms that control action potentials. This link is a small chemical molecule called a *second messenger*. It is like a neurotransmitter, but instead of spanning the distance between cells, second messengers work inside them. Receptors are linked to and control specific enzymes that produce second messengers. The activity of these enzymes is controlled by the binding of the neurotransmitter to the receptor. So, the process of binding can turn on an enzyme that makes the second messengers, which diffuse throughout the cell and help to regulate the firing of action potentials.

In addition to controlling the rate at which the cell generates action potentials, second messengers can also help to control how the cell responds to its local environment. Second messengers may even influence the activation of certain genes. Because genes are the blueprints for all the components of the cell, turning genes on or off determines how many and which kind of components the cell will produce. If a cell needs more receptors, for example, the gene that contains the blueprint for that kind of receptor can by turned on by a second messenger.

Let's look at how second messengers work in regulating action potentials. Recall in the boats-and-bikes model that it took $10.00 to hire either a motorcycle or a traffic cop, but each captain only has $1.00 aboard. But let's say that no individual captain can leave his boat to gather all the money together. The captains need an agent working on the pier to do this for them. This agent, who runs from boat to boat collecting money from the captains to give to the bikers or to the traffic cops, is like a second messenger. In the next chapter, we will see how important second messengers are for the actions of some drugs of abuse.

REUPTAKE

Once neurotransmitters bind to receptors, either to elicit action potentials or prevent them from firing, their job is done. The receptors release the neurotransmitters, which quickly move back into the synapse. Some kinds of neurotransmitters are destroyed by special enzymes in the synapse. Other kinds of neurotransmitters are removed by glial cells, the special supporting cells of the nervous system. Still other kinds of neurotransmitters are gathered up and returned to the neurons that originally released them. This gathering up process is called reuptake. The structures that gather up the neurotransmitters are called transporters. The transporters are large molecules that span the cell membrane of the axon terminal, just as receptors span the cell membrane of dendrites. The transporter binds to neurotransmitters in the synapse and deposits them back inside the axon terminal of the neuron that released them. There, like shipyards that haul in and dry-dock ships for the winter, the vesicles can gather and store these neurotransmitters until the neuron needs to send them out again. (Reuptake is discussed in more detail in Chapter 5.)

ADDICTIVE DRUGS DISRUPT NEUROTRANSMISSION

Addictive drugs change the brain's communication system by interfering with synaptic transmission. Some drugs mimic certain neurotransmitters and convey false messages. Other drugs block neurotransmitters and prevent real messages from getting through. Still other drugs have different kinds of effects that modify the flow of information among neurons. But *all addictive drugs interfere with the way neurons communicate*. They change the way the brain works, and *that* changes how people perceive the world, how they feel about themselves and their world, and how they behave.

CHAPTER 5

HOW ADDICTIVE DRUGS CHANGE
THE WAY NEURONS COMMUNICATE

*B*efore we look at how various drugs change the brain's messages, let's
see what can happen when they do.

Allison and Her Friends

"Come on, Allie, say yes!"

Jennifer and Megan have been worried about Allison since she
returned from her grandparents' summer home a few weeks ago.
Allison is depressed and doesn't want to go out with her friends
because her new boyfriend Barry hasn't called once and won't return
her calls. She just isn't her usual, outgoing, fun-to-be-with self. Even
the brand new sports car Allison's parents gave her for her 16th birthday last week hasn't aroused much enthusiasm. Allison's friends
are determined to get her back into circulation and cheer her up.

"Todd's parents left for Europe yesterday," Jennifer says, "and
all the kids are going to a party at his house tomorrow night after
the football game. It'll be great; no parents, no hassle, and every-
body will be there. You have to come, Allison."

"Besides," adds Megan, "you're the only one who can drive us
there! Come on, Allie, say yes! Pleeeease?"

"Oh, all right." Allison gives in. "Maybe it'll be fun. What time
should I pick you up . . . ?"

The next night, the team wins the game, everyone goes to
Todd's, and, after a couple of hours of partying, everyone, including Allison, is in a great mood.

About 100 kids show up at Todd's house. There is lots of great
music, several kegs of beer, and plenty of drugs. Everyone is
having a good time. While Allison and Jennifer flirt with a couple
of guys from the team, Todd approaches.

"So, Allison, I hear you've got cool new wheels," Todd says. "How about taking me for a quick ride?"

"Okay. You drive," Allison says, tossing Todd the keys.

"I'm coming, too," says Jennifer.

"Me too," echoes Megan. As they all pile inside, laughing at how little room there is, other friends gather around to admire Allison's new car.

Someone hands Todd another beer. He chugs it as he turns on the ignition and waves goodbye.

"Back in a sec," they yell, as they speed down the driveway and out onto the road.

...But they aren't back in a second. The carefree, but drunk and stoned teenagers, laughing and singing, enjoying the thrill of testing just how fast the car can go, miss a turn and pierce the night with the sounds of brakes squealing, screams, metal ripping apart, glass splintering, and then horrifying, stunning silence.

The next morning's headline says it all: TEENS' CAR HITS TREE, FOUR DEAD.

Although this episode is a rather dramatic example, tragedies such as this are a far-too-common occurrence among American teenagers who combine driving with drugs or alcohol. In fact, drunk and drugged driving deaths among teenagers generated a public health crisis in the 1970s. The unprecedented epidemic of illicit drug and alcohol use then killed so many adolescents that their life span actually decreased, whereas that of all other age groups lengthened. This crisis generated a major prevention effort on the part of American parents, which contributed to a significant reduction in drunk-driving deaths, cut regular drug use in half among all ages, and reduced it by two thirds among adolescents and young adults throughout the 1980s and early 1990s. Alarmingly, national surveys show that both drunk-driving deaths and drug use among adolescents and young adults began to rise again in the early 1990s.

The reason Todd thought it was all right to drink and drive is the direct, acute result of the way alcohol changes the brain's communications system. Todd knew that drinking and driving don't mix. And when he wasn't drinking, he could convincingly present all the reasons why he would not engage in such behavior. In fact, Todd, Allison, Jennifer, and Megan all had signed school pledges that they would not drink and drive. So, why did they?

When people start to drink, alcohol impairs their judgment. We can see this clearly with these teenagers. None of them realized they were drunk, they just knew they were having a good time. But the amounts of alcohol they drank at Todd's party impaired their judgment enough to get in the car and head down the road without asking themselves whether they should. They paid a tragic price for ignoring that question.

The following sections explain what neuroscientists know about how alcohol, nicotine, and other drugs that people abuse change the brain during intoxication and over time. The text is divided into broad categories with discussions about the commonly abused drugs in each category. Drugs are categorized in a number of different ways. Some categories — agonists and antagonists — are defined by what drugs do when they bind to receptors. Other categories — depressants and stimulants — are defined by their overall effect on the nervous system. Still other categories — designer drugs — are defined by how drugs are made. The category of inhalants is defined by how people take them. Finally, there is phencyclidine, which has such complex actions it does not fit conveniently into any category.

WHY DRUGS CAN CHANGE THE WAY NEURONS COMMUNICATE

Synaptic transmission, which allows the brain to process information with great precision, flexibility, and subtlety, also leaves it vulnerable to drugs of abuse. That's because drugs of abuse, in effect, masquerade as neurotransmitters and interact with receptors and other components of the synapse. Drugs therefore interfere with normal synaptic transmission by introducing false messages or by changing the strength of real ones. Disrupting the transmission of information at the synapse is the basic mechanism by which drugs change behavior. Let's explore this process.

On first examination, undertaking such an exploration may seem overwhelming. Many of the drugs people abuse have a wide variety of effects. For example, the description of all the effects that opiates exert on the brain and body occupies several pages of very small print in a comprehensive pharmacology textbook. Nicotine and alcohol are just as complex. You might therefore assume that it takes dozens and dozens of different kinds of neurotransmitters and receptors to produce this large assortment of drug effects. But it does not. The diversity of a drug's actions is not necessarily matched by a similar diversity in its synaptic effects. Moreover, the number of specific actions a drug exerts can be entirely unrelated to the number of different kinds of receptors with which the drug interacts. In fact, we need only refer to about a dozen different kinds of receptors to explain almost all of the effects of the commonly abused drugs. Some drugs with numerous effects, such as opiates or nicotine, each exert their effects on just one basic kind of receptor.

How can this be? The answer is simple: *localization of function*. For example, many parts of the nervous system contain opiate receptors. And, even though an opiate such as heroin has the same effect on opiate receptors wherever they are found in the body, the neurons that make these opiate receptors have different functions in different locations. There are neurons in the intestines, brain stem, and nucleus accumbens, for example, that each contain opiate receptors. The neurons in the intestines look and work just like the neurons in the brain and control the movements of the intestines. Some brain-stem neurons with opiate receptors control breathing and coughing. Neurons with opiate receptors in the nucleus accumbens regulate feelings of pleasure and reward. Opiates affect all these different kinds of functions.

THE BRAIN REWARD SYSTEM — THE *REAL* DRUG SCENE

We will not try to describe every effect of every drug. Instead, we will use a small number of well-known or obvious effects to highlight some of the key points about each drug. In every case, we will focus on how the drug turns on the brain reward system. We discussed the brain reward system briefly in Chapters 3 and 4. Now, let's enhance that description.

A group of dopamine-releasing neurons, called the *ventral tegmental area* (VTA), forms the core of the brain reward system (see Chapter 3, Fig. 3.4). These neurons send their axons to a variety of limbic system structures, including the nucleus accumbens. Taken as a whole, the circuitry of the brain reward system is complex. One scientist described a drawing of this circuit as looking like a map of the New York City subway system. For our purposes, two structures in this system are most important: the VTA and the nucleus accumbens.

Enough evidence has accumulated to indicate that the release of dopamine into synapses within the nucleus accumbens underlies the generation of reward and the feelings of pleasure we identify with it. Because this single event, the release of dopamine into the nucleus accumbens, is so critical to understanding drug abuse and addiction, we will focus on how each drug of abuse affects the amount of dopamine within the nucleus accumbens. Some drugs increase dopamine by acting on the nucleus accumbens directly. Other drugs increase dopamine levels in the nucleus accumbens by exciting the cell bodies of the dopamine-containing neurons in the VTA. As the VTA cells fire more action potentials, they release more dopamine into the nucleus accumbens. Still others have actions at both sites.

Scientists understand neither why the release of dopamine is so important nor what the neurons in the nucleus accumbens do in response to dopamine to generate reward. They do know that if something eliminates or blocks dopamine or if something damages the nucleus accumbens, the rewarding effects of drugs disappear.

We will focus on the increase in the amount of dopamine in the synapses in the nucleus accumbens because almost all drugs of abuse share this mechanism of action. We do not mean to indicate that other things do not happen, only that the release of dopamine is key.

AGONISTS

An *agonist* is a chemical that binds to a specific receptor and produces a response, such as excitation or inhibition of action potentials. Opiates, cannabis, nicotine, and hallucinogens such as LSD are agonists.

Opiates

Opiates, one of the most studied and best understood classes of drugs, come from the seed pod of the opium poppy. The opiate most abused in America is heroin, but there are many other kinds of opiates, including opium, codeine, morphine, hydromorphone (Dilaudid), methadone, and meperidine (Demerol). The first three can be purified directly from the opium poppy; the last three are made by chemists in laboratories. Note that almost all the latter drugs are medications that doctors prescribe. They too are abused some of the time. How a drug can be both a useful medication and a drug of abuse confuses most people. We will address this paradox in Chapter 11, when we explain how some drugs that produce addiction can also be used safely as medicines.

Opiate-like drugs behave like inhibitory neurotransmitters; their effects are similar to the effects of GABA. As we explained in the previous chapter, this means that opiates attach themselves to, or bind to, specific kinds of receptors — in this case, opiate receptors — which work to prevent neurons from firing action potentials.

Scientists directly demonstrated the existence of opiate receptors in the early 1970s. It didn't seem to make much sense for the brain to make receptors that recognize chemicals from a plant, so scientists reasoned that the brain must contain its own chemicals that the opiate receptor also recognizes. Following this logic, scientists soon discovered three types of opiate-like substances in the brain. But to their surprise, they found that these substances are not chemicals (called alkaloids) like the ones derived from the opium poppy. Instead, they are small pieces of proteins, called *peptides*. Scientists named these opiate-like peptides endorphins, enkephalins, and dynorphins and refer to them generally as *endogenous* (something produced by the brain or the body) *opioids* (opiate-like).

It was the shape of these peptides, not the plant alkaloids, that the opiate receptors were designed to recognize. The fact that these receptors, which evolved in our brains to recognize endogenous peptides, also recognize a plant product turned out to be an evolutionary coincidence. This coincidence is fortuitous because opiates are potent analgesics, but at the same time unfortunate because opiates are potently addicting as well.

Action of Opiate Receptors

Many parts of the brain and the rest of the body contain opiate receptors. But no matter where these receptors are found, they all allow endogenous opioids, as well as opiates, to do the same thing: *slow down or inhibit the activity of the neurons that contain them.* For example, the intestines contain neurons that control the movement of the intestinal muscles. When these muscles are overactive, diarrhea results. Because the neurons that control these muscles contain opiate receptors, opiates can treat diarrhea by reducing the number of action potentials these neurons produce, which in turn slows down the movements of the intestinal muscles. Some of the earliest medications used by man contained opiates to treat diarrhea.

Several kinds of neurons in the brain stem also contain opiate receptors. In one part of the brain stem, there are opiate receptors on neurons that control coughing. At one time, the most effective cough medicines contained codeine. (Some still do, but, because of the abuse liability of codeine, it has been replaced by other kinds of drugs that people do not abuse.) Codeine, a relatively weak, but effective, opiate binds to these receptors, reduces the number of action potentials fired by these neurons, and thereby inhibits coughing.

Effects on Breathing

Another set of brain-stem neurons controls respiration. These neurons also contain opiate receptors, which explains why an overdose of heroin nearly killed Henry (see Chapter 2). The overdose supplied so much heroin to the neurons in the respiratory center of his brain that they became completely inhibited and insensitive to the excitatory inputs (messages) that normally make Henry breathe. As a result, the muscles that controlled his lungs got no commands to contract. His lungs became still, and Henry nearly died.

Heroin stopped Henry's breathing because, as an opiate agonist, it inhibited the brain-stem neurons that control breathing. The emergency room doctor administered an opiate antagonist called naloxone. It replaced the heroin (actually the body breaks down heroin into morphine — more on that in Chapter 6) at opiate receptors all over Henry's brain and body. By replacing morphine on the brain-stem neurons that control respiration, naloxone allowed the normal excitatory processes that make people breathe to resume, and Henry started breathing again right away.

Henry's experience is a dramatic example of an opiate effect. Naturally occurring, endogenous opioids modulate breathing in more subtle ways. In fact, the brain's neurons are commonly exposed to a constant stream of excitatory and inhibitory inputs, which wrestle to control the neuron's activity. If the excitatory inputs become dominant, the neurons fire more action potentials. If the inhibitory inputs become dominant, the neurons cease firing. With breathing, the rate and depth of respiration can vary greatly, depending on the immediate needs of the body. The neurons that control breathing receive various kinds of information, including how much carbon dioxide and oxygen are in the blood, from several brain regions. As the level of carbon dioxide increases and the level of oxygen decreases, the brain sends more excitatory inputs to the respiratory neurons. The sum of all this information determines how fast and deep we breathe.

Whether people take opiate drugs for medicine or to abuse them, they introduce a much greater quantity of opiates into the brain than it is accustomed to dealing with. That is why profound effects, like the complete cessation of breathing, can occur. The subtle effects are going on all the time, but when people take opiates as medicines, or to abuse them, opiates can overwhelm the system. This explains why doctors prescribe carefully regulated doses of drugs such as morphine, meperidine (Demerol), or hydromorphone (Dilaudid). The correct dose relieves pain and aids healing. Too little of the drug fails to do this, and too much can depress respiration (although this just doesn't happen with proper medical use).

How Opiates Produce Reward

Many other parts of the brain, including the brain reward system, contain opiate receptors, which explains why opiates are rewarding. How opiates elicit these feelings of pleasure is complex, but provides a fascinating example of how the brain works.

Opiates produce reward through two separate mechanisms that act in combination.

- They increase the activity of the dopamine-containing neurons of the VTA.

- They act on the neurons in the nucleus accumbens that receive the dopamine message.

Let's start in the VTA (Fig. 5.1). It is not surprising that the dopamine-containing neurons of the VTA are under tight control. Like the neurons that control breathing (and most other neurons all over the brain), they receive both excitatory and inhibitory inputs. The sum of these inputs determines whether and how fast these dopamine neurons will fire.

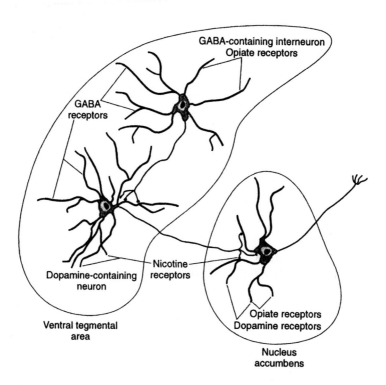

Figure 5.1 Opiates, alcohol, and nicotine produce reward because their receptors are located on key neurons in the brain reward system. As shown, dopamine-containing neurons of the ventral tegmental area contain nicotine receptors, which excite these neurons to release more dopamine. The ventral tegmental area also holds GABA-containing interneurons. Opiates and alcohol act through these receptors to inhibit the GABA neurons, thereby allowing the dopamine neurons to become more excited.

A group of neurons that helps regulate the dopamine neurons of the VTA contains the inhibitory neurotransmitter, *GABA* (gamma-aminobutyric acid). When GABA neurons are active, they inhibit dopamine neurons, which therefore release less dopamine. It turns out that these GABA neurons contain opiate receptors, and when opiates bind to these receptors, they prevent the GABA neurons from firing action potentials. Therefore, the GABA neurons send fewer inhibitory signals to the dopamine neurons, and the latter generate more action potentials to release more dopamine from their axon terminals in the nucleus accumbens.

The release of dopamine is the key event that produces pleasure and reward. So, anything that inhibits the neurons that inhibit the dopamine neurons leads to the release of more dopamine and intensifies feelings of pleasure and reward. That is precisely what opiates do.

In addition, the neurons of the nucleus accumbens that are across the synapse from the axon terminals of the dopamine-releasing neurons also contain opiate receptors. As a result, opiates produce reward by acting directly on the nucleus accumbens. Remarkably, it appears that dopamine receptors and opiate receptors are linked to the same second-messenger system. This means that they produce the same effect on the nucleus accumbens neurons. Because of this, the rewarding effects of opiates are similar to the rewarding effects of dopamine. This may be one reason why opiates produce such powerful feelings of pleasure.

Cannabis (Marijuana)

Opiates change the way the brain communicates because their chemical *shapes* resemble natural substances in the brain. Thus, receptors designed to recognize endogenous opioids recognize plant-derived opiates as well. The same principle applies to marijuana, although we don't understand as much about how it works as we do about how opiates work. Marijuana comes from the leaves of a plant — in this case the cannabis plant. It contains many different chemicals. The major active ingredient in marijuana, tetrahydrocannabinol (THC), produces the drug's psychoactive effects. Like opiates, THC binds to its own receptor, called the *cannabinoid receptor*, which is widely and densely distributed throughout the brain.

Scientists recently identified the endogenous neurotransmitter that activates cannabinoid receptors. They named it *anandamide*. We do not yet know how anandamide and cannabinoid receptors participate in normal brain function, but we do know that THC can activate the brain reward system and increase dopamine release in the nucleus accumbens. Although it does not produce a particularly powerful effect, its ability to release dopamine certainly helps to account for its widespread abuse.

Effects on Short-Term Memory

In addition to activating the brain reward system, marijuana also disrupts short-term memory. *Short-term memory*, also called *working memory*, allows you to take in information coming from your senses, from other parts of your brain, and from long-term memory and to hold that information in your consciousness long enough to think about it. It is the information in your working memory that constitutes your awareness and defines your consciousness. You use your short-term memory when you look up a telephone number to make a phone call. Even if you have never seen the number before, you can usually remember it long enough to punch it into the telephone after having seen it only briefly. Your ability to carry out short-term memory tasks depends on a properly working *hippocampus*. And a key part of the hippocampus, called the *dentate gyrus*, has one of the highest concentrations of cannabinoid receptors in the entire brain.

Your hippocampus receives information from all parts of your brain. All this information — whether it comes from the outside world through your senses, from the parts of your brain where you store memories, or from the parts of your limbic system that produce emotions — reaches your hippocampus through the dentate gyrus, which acts as a relay station for the rest of the hippocampus. This relay station is loaded with cannabinoid receptors. Neuroscientists believe that marijuana disrupts short-term memory by activating these cannabinoid receptors at the entrance to the hippocampus. This disruption can have profound effects.

On the most trivial level, people high on marijuana have a hard time conducting a coherent conversation. The more complex the subject, the more difficult it is. And, if they have to deal with a complex situation in the real world, it can become a significant problem.

The inability to hold thoughts in working memory makes it difficult (at best) to have good judgment. This also makes it difficult to learn effectively, because thoughts that can't be held in working memory also can't be stored in long-term memory. This cognitive deficit, combined with other effects of marijuana such as the distortion of time, makes complex physical tasks like driving risky activities. Remember Neil, who ended up in a ditch, embarrassed, but not hurt. He was lucky, but he could just as easily have crossed the road and hit an oncoming car.

Anandamide's Role

Because of the many personal and social problems that marijuana creates, the drug has attracted a good deal of scientific interest. In the course of trying to uncover the neurobiological basis for marijuana's actions, neuroscientists discovered a unique window through which to study the brain. Marijuana's ability to disrupt short-term memory suggests a normal physiological role for anandamide. Perhaps its action on cannabinoid receptors in the hippocampus eliminates unneeded information from working memory, allowing new information to take its place. Although this role for anandamide is only speculation, trying to understand how drugs work generates experiments that may help us further understand how the normal brain works. Other drugs of abuse have opened other windows of discovery. The opiates are perhaps the best example of how research on drugs has enriched our understanding of normal brain functions. The discovery of endogenous opioids had a special role in increasing our understanding of how the brain processes pain messages.

Coincidentally, both cannabinoid and opiate receptors work through the same second-messenger system, although opiates turn this system on much more potently. Cannabinoid receptors do not appear to be as strongly linked to this second-messenger system as opiate receptors. Using our boats-and-bikes model, it's as if all the opiate boats land in docks at piers that have lots of agents to collect their money, but many of the THC boats land in docks at piers that have no agents. Thus, every time an opiate boat lands, it can contribute something to the message. In contrast, even though they land, most of the THC boats cannot contribute anything at all.

All neurons probably contain the second-messenger system that both opiates and cannabinoids rely on. Opiate receptors and cannabinoid receptors, however, are not usually found on the same neurons. This explains why these drugs have different effects, even though they act through the same second-messenger system.

Other Effects

In addition to altering mood and memory, marijuana also has many other effects, such as the following:

- It diminishes coordination.
- It impairs the ability to carry out mental tasks that require several sequential steps of reasoning.
- It makes time appear to slow.
- It distorts perceptions of the world (sometimes the senses seem to merge together).
- It can make users feel as if they are not really part of themselves.

These effects all disappear as the drug "wears off." Some of the effects of marijuana use can even reverse themselves while the user is intoxicated. Right after smoking, for example, users may feel mild euphoria and some excitation and increased energy. Not too long afterward, the opposite effects — sedation and sleepiness — appear. All these effects are complex, and we do not yet understand the brain mechanisms that underlie them.

Nicotine

Nicotine is another drug that mimics a naturally occurring neurotransmitter. It acts on a special type of receptor that recognizes a neurotransmitter called *acetylcholine*. (Acetylcholine was the first neurotransmitter scientists discovered.) Neurons in your spinal cord use this transmitter to control your muscles. Scientists call the special neurons that control muscles *motor neurons*. The cell bodies of motor neurons reside in the brain stem and spinal cord, and their axons terminate on muscles.

Motor neurons constitute the final common pathway over which the commands that control our movements reach our muscles. When the axon terminals of motor neurons release acetylcholine, muscles contract. The more acetylcholine they release, the more forcefully the muscles contract. If nicotine is placed in the synapse between the terminals of a motor neuron and a muscle, nicotine will also make the muscle contract because it mimics acetylcholine.

Multiple Effects

There are two kinds of acetylcholine receptors, but only one responds to nicotine. This one is called the *nicotinic cholinergic receptor* ("cholinergic" is the adjective derived from "acetylcholine"). Nicotinic cholinergic receptors are found in a variety of places within the brain and body, but, unlike opiate and cannabinoid receptors, which *inhibit* neurons, nicotinic receptors *excite* neurons to fire action potentials. Nicotine receptors appear not only on dendrites and cell bodies, but also on axon terminals. When nicotine binds to its receptors on dendrites and cell bodies, it initiates action potentials that eventually cause axon terminals to release neurotransmitters. But when nicotine binds to the receptors on axon terminals, the terminals release even more transmitter.

Nicotine produces feelings of pleasure and reward by acting on both cell bodies and axon terminals. First, the cell bodies and dendrites of the dopamine-containing neurons of the VTA (ventral tegmental area) contain nicotine receptors. So, when Sybil smokes a cigarette, the nicotine she inhales directly excites the key dopamine-containing neurons of her brain reward system. In addition, nicotinic receptors reside on the axon terminals of the dopamine-containing neurons. Nicotine not only excites these neurons, but when they actually fire an action potential, nicotine also increases the amount of dopamine they release.

Nicotine does not act exclusively on the brain reward system. It has other effects that many people find desirable as well. Nicotine can reduce anxiety, improve attention, and relax muscles. It also suppresses appetite. As with the opiates, the location of nicotinic receptors on neurons in different parts of the brain largely accounts for the drug's many different effects.

Hallucinogens

Hallucinogens are a diverse group of drugs with the wrong name: they do not produce hallucinations. Rather, they *alter* perception, thought, and feeling. Strictly speaking, hallucinations are perceptions of things that do not exist. Hallucinogenic drugs *transform* perceptions of things that do exist. To make this distinction explicit, the term *psychedelic* was introduced. Unfortunately, this term accumulated negative connotations by being associated with social and political movements that surrounded drug use in the 1960s and 1970s and never came into widespread use. Whatever they are called, these drugs have profound effects on the brain. People who have used LSD or mescaline, for example, have reported that the different senses may merge so that sounds are "seen" or colors "heard." Users have also reported that sensations may be heightened and seem more clear or immediate. Common objects or events can become fascinating, as if they were very special and had never been experienced before. Or, sensations may be distorted, so that straight lines, for example, appear wavy. Many users describe an LSD "trip" as dream-like.

Emotions are affected as well. As a result, people high on hallucinogens may cry or laugh inappropriately, or believe that they have had profound insights into themselves, their relationships with other people, or some aspect of the world. Once the drug has worn off, however, the importance of these insights starts to wane.

There are several different kinds of hallucinogens. One class includes drugs such as LSD (acid), mescaline, and psilocybin (magic mushrooms). These drugs act directly on a fourth class of receptors, called *serotonin receptors. Serotonin* is a neurotransmitter found in many of the same parts of the brain as is dopamine. It is found in many other parts of the brain as well, especially the cerebral cortex. Serotonin is involved in a remarkably wide range of brain activities, including global functions like mood and basic survival functions like sleeping, eating, and sexual responsiveness. It also appears to play a role in determining how incoming sensory information is perceived and how information from different senses is integrated so we have a coherent picture of our whole environment.

Role of Serotonin Receptors

Rather than just one, there are at least 10 different kinds of serotonin receptors. This partly explains why scientists have found it difficult to learn precisely how hallucinogens work. LSD, the best-studied drug of this class of hallucinogens, acts as an agonist at some of these receptors and as an antagonist at others. Because of this complexity, the only thing we can say with any confidence is that LSD and related hallucinogens disrupt the way serotonin normally works in the brain.

Apparently, the actions of hallucinogens on serotonin receptors account for their psychedelic effects. But it is not at all clear how the heightening, distortion, and blending of the senses, or the increased emotionality that users experience, are directly related to the activation or blockade of serotonin receptors.

Animals Say "No"

We also do not understand another aspect of hallucinogen abuse. Although people abuse hallucinogens, these constitute the only class of drugs of abuse that animals will not self-administer. Hallucinogens do not appear to activate the brain reward system. This makes one wonder why people abuse them. One suggestion is that human users of hallucinogens may be trying to escape boredom with sensation-seeking — something survival-oriented animals would almost certainly avoid.

It is interesting that the new antidepressant medications — fluoxetine (Prozac) and sertraline (Zoloft) — are thought to work by increasing serotonin levels in the synapse. The difference in the effects of these antidepressants and hallucinogens is just one indication of how complex the effects of serotonin can be. We will learn even more about serotonin when we examine alcohol's mechanisms of action below.

Marijuana, PCP (angel dust; phencyclidine), and MDMA (methylenedioxymethamphetamine) are also sometimes classified as hallucinogens. It is interesting that the hallucinogenic effects of marijuana and PCP do not appear to depend on interactions with serotonin. Like LSD, MDMA also interacts with serotonin receptors, but has stimulant properties as well (see below).

ANTAGONISTS

Other kinds of drugs, called *antagonists*, block receptors in the brain. They actually bind to the receptor, but produce no response. In addition, antagonists can prevent agonists from reaching receptors to produce their response. A common antagonist is caffeine. A less common one is PCP.

Caffeine

The most widely used drug in the world is caffeine. People have consumed products from plants that contain caffeine (found in coffee and many carbonated beverages) and its chemical cousins, theophylline (found in tea) and theobromine (found in tea and cocoa), for many thousands of years. In the doses that most people consume, these drugs, as mild stimulants, produce slight elevation in mood and increased alertness. Moreover, in fatigued people, these drugs increase the ability to concentrate and work. High doses of caffeine and caffeine-related substances can lead to anxiety and sleeplessness.

Caffeine exerts its effects on the brain primarily by interacting with a specific class of receptors, called *adenosine receptors*. Unlike the drugs previously discussed, caffeine, as an antagonist, blocks the receptor to which it binds. This prevents the endogenous neurotransmitter (adenosine) from binding to the receptor and exerting its effect.

Theophylline, a caffeine-like drug, has long been an important medicine in treating asthma. As a result, we know a great deal about how such drugs work to relax smooth muscles, such as the muscles found in the airways of the lungs. In contrast, we know much less about caffeine's effects on the brain, largely because these effects are relatively weak and because caffeine is widely accepted as a benign drug.

Like other drugs of abuse, caffeine is rewarding, and it can produce tolerance and dependence. Still, many people don't consider it addictive because the consequences of its use are so mild when compared with drugs like nicotine, alcohol, and heroin. Because caffeine produces few, if any, personal or social consequences, scientists have not been particularly motivated to study its effects on the nervous system in great detail. Most believe that it produces its effects because of the locations of the neurons that contain adenosine receptors, but little clear evidence supports this belief.

DEPRESSANTS

Depressants are drugs that relieve anxiety and produce sleep. The depressants we will discuss first are synthetic products of chemistry laboratories. A
synthetic drug can be an exact copy of a drug found in nature that chemists
make using pure chemicals, or it can be a chemical that was created for the
first time in the laboratory. Even though they are synthetics, the depressants
described below act much like alcohol in relieving anxiety and producing
sleep. We will finish this section with a discussion of alcohol.

Depressants generally act to reduce action potentials. They literally depress
the nervous system. Depressant drugs include barbiturates, benzodiazepines,
and alcohol.

Barbiturates and Benzodiazepines

Depressants are not directly toxic to cells as alcohol is, and, although they all
have potential for abuse, they have important medical uses as well. Common
depressants are tranquilizers such as diazepam (Valium), alprazolam (Xanax),
and sleeping pills such as flurazepam (Dalmane), triazolam (Halcion),
pentobarbital (Nembutal), and secobarbital (Seconal). Valium, Xanax, Dalmane, and Halcion belong to a class of drugs called *benzodiazepines*, whereas Nembutal and Seconal are members of another family, called *barbiturates*.
Benzodiazepines are safer than barbiturates and therefore have replaced barbiturates for most medical uses. Moreover, people are less likely to abuse
benzodiazepines, although they will and do.

Depressants Act on GABA

All depressant drugs work by potentiating the action of the inhibitory neurotransmitter GABA. These drugs cannot activate the GABA receptors by
themselves and do not bind to the same site that GABA does; they bind to
their own places on the GABA receptor. Once they have bound to this site,
any GABA that binds to its receptor has a much more powerful effect.
Because GABA receptors reside all over the nervous system, depressant
drugs taken in high enough doses can diminish the ability of almost all parts
of the brain to function properly. Taken at therapeutic doses, they primarily
reduce anxiety and arousal.

Although both benzodiazepines and barbiturates share the potentiating
action at the GABA receptor, barbiturates have additional effects, which
they share with alcohol. Most important, they *inhibit excitatory transmission*
at the same time that they are potentiating inhibitory transmission. This

effect is due to an action on glutamate receptors. Because of this additional action, higher doses of barbiturates can produce surgical-level anesthesia, coma, and death (by stopping respiration).

Barbiturate overdose has caused many accidental deaths in abusers, and many people have used barbiturates to commit suicide. Benzodiazepines alone, even at very high doses, rarely kill, because they do not suppress excitatory transmission the way alcohol and barbiturates do. Because they share mechanisms of action, it is not surprising that both benzodiazepines and barbiturates enhance the actions of alcohol (and vice versa). Combined with alcohol, either benzodiazepines or barbiturates form a dangerous and potentially lethal mixture.

Depressants Are Also Rewarding

Both benzodiazepines and barbiturates appear to activate the brain reward system by acting at the inhibitory neurons in the VTA. (Alcohol has the same action on these neurons.) But barbiturates produce a much more potent reward than either alcohol or the benzodiazepines. The depressant drugs are also reinforcing for a second reason. They relieve anxiety and stress. So, even though benzodiazepines do not produce feelings of euphoria, many people abuse them because relief from anxiety is itself reinforcing. We describe this mechanism of reinforcement in more detail in Chapter 7.

Alcohol

Alcohol is a very simple chemical, but a remarkably complex drug. Like caffeine and opiates, people have used alcohol for thousands of years. Its use is deeply embedded in both social custom and religious ritual. Many people would agree that when used within the constraints of these customs and rituals, alcohol has real social benefits. Moreover, considerable evidence now exists that modest use of at least some alcoholic beverages, wine especially, may have health benefits by reducing heart disease (but it may increase the risk for breast cancer).

Regardless of its potential benefits, alcohol has a high potential for abuse. Alcohol is a toxic substance that can directly damage cells, and prolonged exposure to high doses damages the brain, liver, stomach, and other organs. When people use it improperly, alcohol's effects on personal health and the social fabric can be devastating. These effects can be acute and immediate, as we have seen with the alcohol-related deaths of Todd, Allison, Jennifer, and Megan, and they can be long-term and far-reaching. Almost everyone knows someone who has an alcohol problem or is addicted to alcohol, and has felt the impact of this addiction on his or her own life and the lives of other family members.

Alcohol not only has complex interactions with people and their social and personal health, but it also has complex interactions with the brain and body as well. Alcohol binds to at least three receptors: acting as an agonist at two, the GABA and serotonin receptors, and an antagonist at the glutamate receptor. In addition, it directly influences cellular metabolism in at least four different ways. At higher doses, alcohol can even undermine the structural integrity of the membrane that surrounds each neuron. That's some package!

It is not surprising that scientists have had a great deal of interest in alcohol and have studied its mechanisms of action for some time. However, the ways in which alcohol interacts with brain receptor systems have only recently begun to emerge. Understanding alcohol's effects on these receptors not only explains many of its specific effects, but also helps to explain why alcohol has effects at higher doses that are not seen at lower doses and why these effects change during the course of a single episode of drinking.

Effect on GABA Receptors

Alcohol acts like an agonist at GABA receptors, and it is this effect that is most prominent when people drink small amounts or are at the beginning of a bout of drinking. Like barbiturates and benzodiazepines, alcohol is not really a direct agonist at the GABA receptor, as morphine is at the opiate receptor. That is, alcohol cannot activate the receptor by itself. Instead, it binds to a site on the receptor molecule separate from the GABA binding site (and separate from the barbiturate and benzodiazepine site as well) and potentiates the effect of GABA. It's as if every GABA boat (as in the boats-and-bikes model), which normally has just $1.00 to hire a traffic cop, has $2.00 or $3.00 to hire a cop when alcohol is present.

Alcohol activates the brain reward system by inhibiting the GABA neurons that inhibit the dopamine-containing neurons of the VTA. Opiates act on these same neurons. The difference is that alcohol acts on GABA receptors, whereas opiates act on opiate receptors. In any case, the ultimate effect is the same. Alcohol inhibits the activity of the GABA neurons and thereby takes the brake off the dopamine neurons. These, in turn, become more active and release more dopamine in the nucleus accumbens.

Alcohol's effect on GABA receptors in the cerebral cortex probably accounts for the initial excitement, increased emotionality, and loss of behavioral restraint that occurs soon after a drinker has the first couple of drinks. This explains why Todd and the girls decided it was okay to drive, even though they all knew better. The enhanced GABA effect appears first to disrupt higher cognitive processes, the kind that depend on experience, memory, and practice to perform well. These kinds of cognitive processes underlie accurate perception, good judgment, and smooth social interactions. Todd lost these abilities soon after he started drinking, because these processes depend on information processing that is highly complicated and that requires complex neural circuits with lots of synapses. It is easier to disrupt a process with more synapses (neurons), because it only takes failure at a few synapses for the whole circuit to fail. In other words, by enhancing inhibitory inputs, alcohol essentially creates neurons that behave like people who do not pay attention.

In an earlier chapter, we described a dynamic conflict between the cognitively oriented cerebral cortex and the survival-oriented needs of lower brain centers. If you are watching your weight, for example, you won't eat every time you are hungry. If you wish to get along with your neighbors and colleagues, you won't lose control of your emotions, even when you feel provoked. Using such control, you can decide whether or not to eat or whether or not to fight. You rely on your cerebral cortex (and plenty of training and practice) for this ability to restrain your emotional responses and to use reasoned responses instead.

If something diminished the ability of your cerebral cortex to contribute to the kinds of decisions about when to eat and when to fight, each bout of hunger could produce food-seeking and eating, and each provocation could turn into an argument or fight. Therefore, if alcohol weakens or reduces the ability of your cerebral cortex to influence your behavior, as it did with Todd, you will be more likely to engage in inappropriate behaviors. Having had too much to drink, you might even get behind the wheel and drive, as Todd did, or engage in unprotected sex, even though you normally know better.

From Excitation to Sedation

As the drinker consumes more alcohol, the excitation is soon reversed and turns into sedation. As blood alcohol levels increase with continued drinking, GABA action intensifies, and its inhibitory effect begins to depress the entire nervous system, not just the cerebral cortex. Drinkers become uncoordinated and slur their speech because alcohol depresses the motor system. By this time into a bout of drinking, a second kind of receptor is being influenced as well. Like the barbiturates, alcohol acts as an antagonist at the excitatory neurotransmitter *glutamate*. By binding to the glutamate receptor, alcohol diminishes the excitation that glutamate normally elicits when it binds to these receptors.

This indirect effect (that is, altering the potency of the transmitter by binding to a separate place on the receptor) is analogous to the way alcohol enhances GABA action. The alcohol-induced reduction in excitation of neurons adds to the increased inhibition of neurons mediated via GABA receptors and can profoundly reduce neural activity, eventually leading not only to sedation, loss of coordination, and sleep, but also to anesthesia, coma, and death. Action on the GABA receptor alone does not produce these last three effects; they occur only when the effects on glutamate receptors are added to the effects on GABA receptors.

As we saw earlier in this chapter, some sleeping pills (barbiturates) work the same way as alcohol. The combined increase in inhibitory action and decrease in excitatory action depresses the entire nervous system and, with severe overdoses, can fully depress the respiratory system, resulting in death. Although both alcohol and opiates may cause death by acting on the respiratory neurons, they act through different receptors, which inhibit action potentials in different ways. Because they *do* act on different receptors of the same neurons, the effects of alcohol and opiates can add together, making the combination particularly risky.

In addition to its agonist actions at GABA receptors and antagonist actions at glutamate receptors, alcohol acts as an agonist at one of the kinds of serotonin receptors. Some experimental drugs bind to this particular serotonin receptor and activate it (that is, they are agonists), and people report that they feel as if they have been drinking. Antagonists acting at this receptor also seem to reduce some of the effects of alcohol on motor coordination and may reduce the severity of alcohol withdrawal. These observations suggest that it may be possible to create a drug that will alleviate at least some of the symptoms of alcohol intoxication. If they are accurate, however, such a drug could leave us with well-coordinated, drunk people who might even be more dangerous than the uncoordinated ones we have to deal with now.

Effects on Cell Membranes

In higher doses, alcohol can alter cell membranes. These membranes surround every neuron (and every other cell as well) and serve as the neuron's "skin." Because membranes maintain the electrical charge that neurons need to fire action potentials, any alteration in their structure and function can have widespread effects on the brain. This can be a potentially profound effect because neurons all over the brain could stop working in unpredictable combinations. But these effects require higher doses of alcohol than most people can tolerate and therefore are almost certainly not relevant in explaining the commonly seen alcohol effects. The effects mediated by GABA, glutamate, and serotonin receptors are responsible for the symptoms of alcohol intoxication. Alcohol effects have also been reported at opiate and dopamine receptors, on the activity of second-messenger systems, and on several enzymes involved in cellular metabolism. The implications of all these actions are not yet clear, although the disruption of the enzymes involved with cell metabolism may be what causes the damage seen in organs like the liver after long-term, heavy alcohol use.

STIMULANTS

Although stimulants make people more awake and alert, and increase their energy, stimulants do not act to simply increase action potentials. Their actions are not simply the opposite of the actions of depressants, which act to decrease the number of action potentials. Stimulants excite the nervous system by activating the neural systems that normally create and maintain arousal and attention. This class of drugs includes cocaine, the amphetamines (most prominently, methamphetamine, known as meth, ice, or crystal) and a few other drugs, such as methylphenidate (Ritalin), which is now commonly used to treat attention deficit disorder.

Cocaine and the Amphetamines

In low or moderate doses, cocaine and the amphetamines elevate mood, increase feelings of well-being, and increase energy and alertness. In higher doses, especially when they are injected or smoked, which gets them to the brain rapidly, these drugs can produce an intense euphoria that is a potent reward. At still higher doses, they produce anxiety and restlessness. Very high dose levels can also produce a frank psychosis that resembles paranoid schizophrenia. Many addicts experience these effects because they reach very high levels of drug in the blood using large amounts of stimulants for several days during a binge or "run." This extraordinary turnabout, from feeling good to being insane, is even more paradoxical because dopamine creates *all* these effects. The effect seen is due to the amount of dopamine available in the synapse.

At Low Doses: Reward

Let's deal first with the rewarding effects of stimulants, which is why people abuse them.

Cocaine and the amphetamines produce pleasurable feelings by increasing the amount of dopamine in the synapses of the nucleus accumbens. The message that dopamine relays increases in intensity and duration. This increased amount of dopamine in the synapse occurs through mechanisms that don't have direct effects on receptors.

Reuptake

Both cocaine and amphetamines block the reuptake of dopamine by the neurons that originally released it. In addition, amphetamines can also release dopamine into the synapse by acting directly on the axon terminals.

Once neurotransmitters have delivered their messages, something must remove them from the synapse to turn their messages off. Some neurotransmitters, such as acetylcholine, are quickly destroyed by specific enzymes that are in the synapse. Other neurotransmitters, such as dopamine, are not. Instead, they are rapidly removed from the synapse by a special "reuptake pump." This pump, called a *transporter*, straddles the cell membrane of the axon terminals of the dopamine-releasing neurons. It rapidly ferries dopamine molecules out of the synapse and deposits them back inside the axon terminals. Once there, the terminals either destroy or recycle the dopamine.

Cocaine binds to the dopamine transporter in the same way that some other drugs bind to receptors. By binding to the transporter, cocaine blocks its ability to transport dopamine from the synapse. Because this prevents the axon terminals from retrieving dopamine from the synapse, dopamine, now trapped in the synapse, stimulates and restimulates dopamine receptors and produces a powerful reward.

Amphetamines (along with methylphenidate) also block the dopamine transporter to some extent, but this does not appear to be the major way they produce reward. Instead, amphetamines increase the levels of synaptic dopamine in a more direct way. They get inside the axon terminals of the dopamine-containing neurons and release dopamine directly into the synapse.

Because both amphetamines and cocaine increase synaptic levels of dopamine within the nucleus accumbens, they are potent activators of the brain reward system. Indeed, it is easier to teach animals to self-administer cocaine than any other drug.

At High Doses: Psychosis

Some dopamine is a good thing. It has real value to our survival because it teaches us how to get the essentials, such as food, which we must have to survive. In addition, it makes us feel good. Too much dopamine, by contrast, can be a disaster. Consider these two facts: first, the drugs that are best at treating schizophrenia block dopamine receptors; second, if a person with schizophrenia takes a drug such as cocaine, which increases dopamine levels, his or her symptoms worsen. In other words, the very same kind of receptors that must be activated to produce reward have to be blocked to alleviate the hallucinations and thought disorders associated with schizophrenia.

Too much dopamine can even induce schizophrenic behavior in an otherwise normal person. Stimulant addicts often indulge in binges of drug-taking, which last for several days. During this relatively brief period of time, they use large quantities of drug. Many end up with a syndrome that looks like paranoid schizophrenia and includes the hallucinations that are symptomatic of the disease. In most cases, this "psychosis" disappears as soon as drug use stops, proving its close link to high levels of dopamine.

There is more. The brain reward system itself appears to be part of the neural circuit that schizophrenia affects most profoundly. Blocking dopamine receptors in the brain reward system alleviates schizophrenia. Blocking dopamine receptors in other brain systems does not. How can the same neural system that produces feelings of pleasure also be involved in one of the most serious mental illnesses, one symptom of which is an inability to feel pleasure? The short answer is that we don't know.

It is ironic that dopamine can lead us first to pleasure and then to insanity. Once again, the study of drugs of abuse has opened a unique window into the brain for scientists to explore, not only to answer the question of how drugs of abuse work, but even to learn what causes a major mental illness and find a cure for it.

DESIGNER DRUGS

In the same way that legitimate chemists try to make useful drugs to meet the needs of patients, illicit chemists have tried to make drugs to meet the "needs" of drug dealers, abusers, and addicts. Until relatively recently, the Controlled Substances Act, the law designed to control the availability of drugs that people abuse, listed each illegal drug according to its specific chemical components. "Basement" chemists could accommodate dealers by slightly changing the chemical structure of a drug to circumvent the drug control law. As a result, chemical analogues of abused drugs, whose chemical structures were only slightly modified from their illegal precursors, were legal.

To stay one step ahead of the law, a basement chemist might have made new drugs that were illicit without being illegal. These chemists worked not only with opiates and amphetamines, but with hallucinogens as well to create an alphabet soup of compounds with psychedelic properties, such as DMT (dimethyltryptamine), DMA, DOM, MDA (methylenedioxyamphetamine), and MDMA. To put a stop to this activity, Congress amended the Controlled Substances Act to prohibit chemical analogues of illegal drugs as well, and most states have amended their laws along similar lines.

Designer drugs based on hallucinogens act more or less like LSD. Those derived from amphetamines have stimulant properties along with hallucinogenic properties. MDA and MDMA, the two most notable designer drugs, have properties that combine stimulant and hallucinogenic effects.

MTPT and Parkinson's Disease

The designer drugs that were developed from derivatives of opiates all had opiate-like properties. One drug had a particularly devastating, but from the point of view of scientific discovery, remarkable side effect. A few batches of this derivative, called "China white," contained an impurity that turned out to be a neurotoxin (a chemical that kills nerve cells). This neurotoxin, called *MPTP*, kills the neurons that make dopamine. Drug abusers who used China white ended up with brain damage very similar to the damage that occurs in Parkinson's disease. China white isn't made any more, but MPTP as a neurotoxin has turned out to be a hugely important research tool in the field of Parkinson's disease research.

It is now possible to create specific animal models of Parkinson's disease by administering MPTP to animals. Scientists have recently begun to study drugs that might "rescue" neurons damaged by MPTP. This gives hope for similar treatments for Parkinson's and other diseases in which neurons slowly degenerate and die.

MDMA

MDMA (Ecstasy) is a chemically modified amphetamine with psychedelic as well as stimulant properties. It is not as potent a stimulant as methamphetamine, nor as potent a hallucinogen as LSD. In addition, its effects don't last as long as either LSD or methamphetamine. Although MDMA has been widely abused, abusers rarely become addicted to it. MDMA is a good example of a "designer drug."

College students and other young people use MDMA to get high at parties. Unfortunately, this particular designer drug is also neurotoxic. It destroys the axons and axon terminals of the neurons that contain serotonin. This effect appears to be dose-related (i.e., related to how much a person takes) and was first noticed in animals who had been given much larger doses than people usually take. But it is not clear that any dose is safe.

INHALANTS

Most inhalants fall into one of two large classes: organic solvents, such as glue and paint thinner, and anesthetic gases, such as ether and nitrous oxide (laughing gas). Organic solvents often come in liquid form; they evaporate easily and people inhale the vapors. People inhale gases as well. Both classes of drugs activate the brain reward system, although no one understands the mechanism.

Organic Solvents

Intoxication with solvents resembles intoxication with depressants. Initially, and at low doses, solvents can evoke some excitement and increased energy. At higher doses, solvents sedate, perhaps by depressing nervous system function globally. Because organic solvents are highly toxic, most adults do not abuse them. Instead, the group that uses solvents comprises young people, who have few resources and little or no understanding of how neurotoxic solvents are.

Organic solvents are very good at dissolving fatty substances. Every axon has a covering, called a *myelin sheath*, that is made of a special fat. This sheath works much like the protective plastic coating that insulates an electrical cord. Unfortunately, solvents can dissolve the axon's myelin sheath and disrupt communications both within the brain and throughout the rest of the nervous system.

Anesthetic Gases

Ether

Ether is a highly flammable and potentially explosive substance. It was one of the first anesthetics and was widely used in surgical procedures. Ether has been replaced today by less flammable and more effective anesthetics.

Nitrous Oxide

People used both nitrous oxide and ether to get high long before anyone recognized that these drugs had anesthetic properties. Nitrous oxide is a weak anesthetic that does not actually produce unconsciousness, but leaves the user in a kind of twilight zone between waking and sleeping. This state is very much like the initial stage of excitement that depressants produce, but, unlike depressants, nitrous oxide does not replace this state with further sedation and sleep. We do not know how this anesthetic exerts its effects.

One of the real dangers of inhalants comes from the way people abuse them, rather than from their pharmacological properties per se. To increase the amount of vapor in their lungs, abusers often inhale these drugs from closed containers, usually a flexible bag of some kind. It is not uncommon for a person to die under such circumstances, because the gas vapor replaces all the oxygen in the lungs and essentially the person suffocates. In fact, some health care professionals, who have ready access to the drug and sometimes abuse it, have died from nitrous oxide overdoses while inhaling the gas through face masks. They died either because they failed to combine oxygen with the nitrous oxide in the first place, or because the oxygen ran out before the nitrous oxide did. In either case, nitrous oxide caused a loss of awareness, lack of oxygen quickly caused unconsciousness, and they died.

DRUGS WITH COMPLEX EFFECTS

Phencyclidine (PCP)

PCP, a drug initially developed as a veterinary anesthetic, has an assortment of effects. It is an hallucinogen, a stimulant, a sedative, and an analgesic that acts like no other. Doctors briefly used PCP as an anesthetic for humans, because it does not suppress breathing the way other anesthetics and depressants do. It appeared to be especially safe for babies and older people. But PCP turned out to have intolerable side effects. Patients suffered from confusion, delirium, and frightening dreams while recovering from its anesthetic effects. While it was in use, doctors described PCP as a "dissociative" anesthetic because patients did not appear to become unconscious. Their eyes remained open and they made small uncoordinated movements, but they didn't respond to pain. At doses that are not large enough to produce anesthesia, PCP users typically seem to be drunk. They lose their coordination and their speech becomes slurred. They also describe feelings of being outside their own bodies, and they are often drowsy (though they may become agitated) and insensitive to pain.

PCP is a glutamate antagonist, but it does not actually prevent glutamate from binding at the receptor. Instead, it binds at a separate site and prevents the receptor from responding once glutamate has bound to it. As a result, glutamate can no longer excite the neuron to produce action potentials, as if the boats landed at the docks, but had no way of contacting the motor-cyclists. This glutamate-blocking effect accounts for PCP's anesthetic properties. Glutamate is used by many brain circuits, including those that relay pain messages from the body to the brain. Blocking the excitatory effects of glutamate therefore blocks the perception of pain.

Animals will self-administer PCP directly into the nucleus accumbens, which is strong evidence that it exerts its rewarding effects there. Curiously, these effects do not appear to be dependent on dopamine, which means that PCP must be acting on the neurons that receive the dopamine message rather than the ones that send it (VTA neurons), probably by blocking glutamate inputs to the neurons that also receive the dopamine input. PCP binds to a number of other receptors as well, but it is not clear which effects are due to which receptors. The complexity of PCP's various effects makes it difficult not only to understand how PCP works, but also to understand how individual users will respond to it.

PEOPLE USE DRUGS BECAUSE DRUGS ARE REWARDING

With the exception of the LSD-like hallucinogens, all the drugs of abuse we have discussed activate the brain reward system in one way or another.

- Nicotine turns on dopamine neurons *directly*.

- Opiates (such as codeine and morphine), depressants (such as barbiturates), and alcohol turn on dopamine neurons *indirectly* by inhibiting the inhibitory neurons that otherwise keep the dopamine neurons silent (or at least working very slowly).

- Stimulants (such as cocaine and amphetamines) increase dopamine levels in the nucleus accumbens by acting on the axon terminals of dopamine-containing neurons to release or prevent the reuptake of dopamine. MDMA (ecstasy) probably shares this action.

- PCP acts directly on the same nucleus accumbens neurons that receive the dopamine message.

- It is not yet clear how cannabis or the inhalants activate the brain reward system, but whatever the mechanism, with the possible exception of PCP, the net result is an increase in dopamine levels in the synapses of the nucleus accumbens.

It is this common alteration in synaptic transmission that underlies the rewarding effects of abused drugs. When we ask why people abuse drugs — legal or illegal, highly addictive or weakly so — the key reason is because *drugs activate the brain reward system.* Alterations in synaptic transmission at other synapses in other brain regions account for the myriad other effects drugs produce, but they all have in common this fundamental effect on the brain reward system.

CHAPTER 6

HOW DRUGS GET INTO THE BRAIN — AND OUT AGAIN

EVERY DRUG HAS ITS ROUTE

Each drug has one or more routes by which it can get into the body and then into the blood. And, each drug gets into the brain by way of the blood. The blood distributes the drug throughout the brain and body. Even though the drug molecules end up all over the brain and body, they can have an effect only on the neurons and other cells with receptors that recognize them. Nothing happens to cells without the appropriate receptors. So, the psychoactive effects of a particular drug depend on the locations of the neurons with receptors to recognize it and on which brain circuits these neurons are part of.

Dose and Route of Administration: Interaction

Most people understand that as drug users increase the dose of a drug, it will have more powerful effects. That is, the effect is dependent on the dose. For a psychoactive drug to have an effect that users can feel and discern, a specific amount (an effective dose) must reach the brain. If the user takes too little, she can't perceive that anything is happening. If the user takes too much, he may develop symptoms of an overdose, such as getting sick or passing out.

What many people do not understand, however, is that the route of administration (how the drug is taken) interacts with the dose to help determine the intensity of a drug's effects. Taking a few hits off a joint of marijuana is enough to get someone high (the person would perceive a change in mood, perception, and cognition). But, if that person ate the same amount of marijuana, he would feel nothing. This is a simple example of how dose and the route of administration interact to determine the presence and intensity of a drug's effects. And, most drug abusers have learned this. So, they take drugs in certain ways to deliver just the right amount of the drug to the brain at just the right rate. The rest of this chapter explains how route of administration modifies the effects of drugs of abuse.

ROUTES OF ADMINISTRATION

People take drugs of abuse in a number of different ways: they inhale them, inject them, absorb them through the mucous membranes of the nose, mouth, or other bodily orifices (did you ever have a rectal suppository?), or eat or drink them. People take some drugs only one way, that is, by only one specific route of administration. For example, people almost always drink alcohol, and they mostly smoke marijuana. But people take most other drugs by various routes at different times. For example, they can take heroin by almost any route. Most addicts prefer to shoot (inject) heroin into a vein, but they can also drink it, smoke it, snort it into the nose, or inject it under the skin ("skin popping").

No matter how it is taken, heroin produces an array of well-known opiate effects — it diminishes pain, increases feelings of well-being, suppresses coughing, and slows the intestines. Changing the route of administration will produce variations in how quickly heroin produces these effects, how intense the effects are, how long they last, and whether or not heroin produces a rush. Each user will respond slightly differently to a given dose and route of administration, but the same general changes would be apparent in all of them.

PHARMACOKINETICS

Pharmacokinetics is the study of how the body absorbs drugs, how drugs are distributed throughout the body, and how the body finally gets rid of them. Thus, pharmacokinetic processes determine how rapidly a given dose gets into the blood, how much of the drug gets in, and how long it will be around to interact with receptors.

For example, the absorption of cocaine is markedly slower when people snort the drug compared with when they smoke it. To get into the blood, snorted cocaine must first dissolve in the fluid coating of the upper nose, so that the mucous membranes of the nose can absorb it into the blood. This is a relatively slow process because this surface is not designed to absorb things coming into the nose, but rather to warm and add moisture to the air we breathe in and to prevent bits of matter, like dust particles, from entering our lungs.

By contrast, when people smoke cocaine, the drug passes into the blood as a gas through the lungs, which are designed to be highly efficient in transferring gases between the air and the blood. Not only are the lungs designed to absorb gases into the blood, but they have a huge surface area, about the size of a tennis court, with which to do it. The absorptive surface of the nose, by contrast, is at most only a few square inches. Smoking cocaine puts a relatively large dose of the drug into the blood rather quickly. From there, it's only a couple of heartbeats to the brain. As a result, smoked cocaine is much more effective than snorted cocaine in producing euphoria and does so much more rapidly.

THE KEY TO DRUG ACTION IS THE BRAIN

To understand the drug effects that cause and maintain abuse and addiction, the key organ to study is the brain. So, even though most drugs have many effects on other organs of the body, we will focus on how drugs get to the brain and how different routes of administration might speed up or slow down this process.

To get into the brain, drugs must first enter the blood. Once there, the circulatory system eventually carries the drug to all parts of the body. Within this system, however, the brain occupies a privileged position. Even though it weighs only three pounds, about 2% to 3% of body weight, the brain is so busy that it actually consumes about 20% of the oxygen used by the body. As a result of its unquenchable need for oxygen, about 20% of the blood that leaves the heart goes to the brain. Moreover, if the body experiences problems getting blood to all of its organs, such as after an injury, it preserves circulation to the brain by diminishing or shutting down flow to other organs. Evolutionary wisdom has apparently recognized that without our brains there is no "us."

WHY THE ROUTE OF ADMINISTRATION MATTERS

Because so much of the heart's output of blood goes to the brain, any route of administration that deposits a large amount of drug into the blood quickly will promptly get a large portion of that drug to the brain. In contrast, a route that gets the drug into the blood more slowly decreases the rate at which it gets to the brain and, because of the way the body eliminates drugs, will probably also reduce the amount of drug that finally does reach the brain. Why this is so will be described in more detail below. Most drug abusers and addicts have learned the two most effective ways to deliver lots of drug into the brain quickly: (1) inject the drug directly into the bloodstream or (2) smoke it.

Intravenous Injection: One Fast Track

Doctors and nurses routinely inject medicine into veins because this is an extremely rapid, reliable, and efficient way of getting medicines into the blood and distributed throughout the body. Drug users inject drugs for the same reasons. When a heroin addict injects a dose into a vein, a bolus (concentrated mass) of the drug enters the blood in a couple of seconds. Within a few heartbeats, the heroin passes into the right side of the heart, where all the blood returning from the brain and body goes first. A few more heartbeats send the heroin through the lungs, where the blood goes to replenish its oxygen supply, and then back to the heart, this time to the left side. From there, it takes only a couple more beats to get the heroin to the user's brain and body.

One major reason heroin addicts shoot up intravenously is to obtain a "rush," an orgasm-like feeling that seems to emanate from the pit of the stomach. Taking the same dose by mouth or through the skin or nose just won't do it. The presence of a rush after intravenous administration — not by other routes — shows how important the rate at which a drug gets to the brain can be in producing certain effects.

Addicts who overdose and die are sometimes found with the needle still in their veins. Actually, they don't die right away. What happens is that the effect of the injected drug is so fast and potent that the user loses consciousness before completing the injection and removing the needle. The user dies soon after because the heroin has shut down the brain's messages that tell the lung muscles to contract and breathe.

Smoking Drugs: The Other Fast Track

Drug abusers have also found that smoking can put a large amount of a drug into the bloodstream quickly. But, only drugs that can literally be vaporized (i.e., turned into a gas) can be inhaled with good results. Burning the drug itself inactivates the drug. Powder cocaine burns at a lower temperature than the one required to vaporize it; so smoking powdered cocaine produces little or no effect. The drug is burned off before it can be vaporized. Crack cocaine, by contrast, vaporizes at a lower temperature than it burns, so it is ideal for smoking. The difference between the two forms of cocaine is a minor chemical modification that has no effect on the pharmacological actions, but does change the temperature at which the compound will vaporize, turning from a solid into a gas. Crack cocaine, the tetrahydrocannabinol (THC) in marijuana, and the nicotine in tobacco all are examples of drugs that vaporize before they burn and can therefore be smoked effectively.

People can also smoke methamphetamine and heroin. While a portion of these drugs is being vaporized, however, some burns up as well, so the user needs a lot of drug to produce the desired effect. Drug dealers usually "cut" (dilute) drugs like heroin, cocaine, and methamphetamine with inactive additives (such as baking soda or fine sugar) to increase the amount the user seems to be buying. As these drugs have become more prevalent over the last couple of decades, dealers have been using smaller amounts of additives.

In the 1970s, a "bag" of heroin (the common unit of sale) was only about 4% heroin. By the late 1990s, heroin potency had risen dramatically to 40%. To contemporary users, it seems as if the drug they buy is more pure (or more potent). The active drug is no different, there is just more of it in each bag. It is this increase in "purity" that has made the smoking of heroin and methamphetamine cost-effective for users.

Smoking Tobacco: All the Advantages of Smoking Revealed

Drug smokers take advantage of the efficient system evolution has given us for absorbing gases through the lungs. In the lungs, the body can absorb inhaled gases from the air into the bloodstream by allowing them to pass through only two layers of cells. One layer lines the lung itself. The other layer forms the walls of the capillaries whose job is to bring blood to the lungs to pick up oxygen and dispose of carbon dioxide. The lung itself consists of tiny balloon-like air sacks called *alveoli*. Gases pass through the one-cell-thick walls of the alveoli and into the capillaries, which are directly adjacent to them. Many hundreds of millions of these alveoli (they are very small) are in the chest. This is why, as noted earlier, our lungs have a surface for absorption about as large as a tennis court.

During normal breathing, we fill only 25% of our lungs' alveoli with new air. Although she is not particularly aware of this, when Sybil smokes (see Chapter 2), she tries to take advantage of more of the lung's absorptive surface area. She takes deep breaths, expands her lungs, and draws the smoke down into many more of the alveoli, holding her breath to allow time for her blood to absorb the nicotine from all the smoke. She often inhales several times in rapid succession, filling her lungs with a lot of nicotine in a relatively short amount of time so she can increase her nicotine blood levels quickly. From her lungs, it is only a few heartbeats back to the left side of her heart before her blood distributes nicotine to her brain and throughout her body.

Titration of Dose

In addition to getting a lot of drug to the brain quickly, drug users can use smoking a second way. They can *titrate* (adjust) the amount of drug in the blood and keep it relatively constant over long periods of time. Cigarette smokers especially find this advantageous. For example, once Sybil has taken her first couple of nicotine hits in the morning, she usually takes fewer puffs, fills her lungs with less smoke, and slows down her smoking rate. If she smokes slowly instead of rapidly, she can perceive the effect of the drug and stop smoking when she reaches the desired level of effects she seeks. If that level falls too low, she lights up again and takes in what she needs to titrate her blood nicotine to the higher level she prefers.

As you have probably guessed, all cigarette smokers typically adjust the amount of nicotine in their blood by controlling how frequently they smoke, how quickly they smoke each cigarette, and even how much of each cigarette they consume. It doesn't take too long for even a novice smoker to learn to detect the first subtle signs of nicotine withdrawal and interpret those signs as a signal to light up. Most experienced smokers have learned to titrate their nicotine dose to meet their needs. This is doubly important because nicotine is a toxic substance, and too much can produce unpleasant side effects, such as dizziness and nausea. Most novice smokers encounter these side effects when they first start smoking, but become tolerant to them quickly. After just a little practice, they are able to adjust their nicotine dose to obtain the desired effects while avoiding the unpleasant ones.

Marijuana users can regulate their input in a similar fashion. However, the marijuana smoker must be patient because the effects of marijuana are not felt immediately, and it is easy to become very intoxicated by smoking too much too quickly. The ability to titrate doses has been used as an argument for making marijuana available for medical uses. Now, doctors can prescribe an orally administered drug containing a synthetic version of THC (dronabinol, trade name Marinol), the major active ingredient in marijuana, for the nausea that results from cancer chemotherapy and to treat the wasting syndrome of AIDS by increasing appetite. However, marijuana advocates argue that the oral dose is too high, and patients could benefit more from smokable marijuana, because they could titrate their own dose. We will address this issue in more detail in Chapter 11.

Taking Drugs by Mouth

People cannot inject or smoke all drugs. Other routes of administration, though less effective in rapidly creating high blood levels of a drug, nonetheless effectively get drugs into the body. For example, people can take most drugs by mouth. Indeed, at the end of the 19th century when opiates and cocaine were legal, a number of medicines available over the counter or by mail contained opiates and cocaine. One wine even contained cannabis, and some popular drinks contained cocaine (an original ingredient in Coca-Cola).

Physicians usually prefer to give medications in oral form, because this makes dosing easy. The patient can take the drug by herself and will get a specific amount of the drug each time she takes it. Doctors can thereby regulate the amount of drug in the patient's body over long periods of time.

In terms of the number of users, the most popular psychoactive drug taken by mouth is alcohol. A wide variety of tasty (and not so tasty) drinks contain alcohol, and drinking alcohol is part of many complex social rituals. Wine plays an important role in the rituals of several religions, including Catholicism and Judaism. Common social events, like going out to dinner, may be surrounded by rituals that include drinking sweet, tasty cocktails, fine wines, and aged whiskeys. If all drinking were confined to these kinds of situations, alcohol would not be the problem that it is today. Unfortunately, tasty alcoholic concoctions allow us to ingest large quantities of a naturally toxic substance, which by itself tastes really awful. And all around the world, people drink these large quantities and get into trouble because of them.

So, it is evident that taking drugs by mouth is an effective way of getting them to the brain. The journey from the mouth to the brain is more complicated, however, than the trip that begins inside a vein or the lungs.

How Drugs Taken by Mouth Reach the Brain

Drugs people take by mouth are absorbed through the stomach, the intestines, or both. If you have any experience with alcohol or any common pain reliever for a headache, toothache, or similar pain, you know it usually takes at least 10 minutes before you even begin to feel the effects, and it can take much longer if you have food in your stomach and intestines. So, drugs taken by mouth get into the bloodstream relatively slowly. Not only that, once drugs enter the stomach, they are exposed to all the digestive chemicals and enzymes waiting to digest food. Indeed, our stomachs contain *alcohol dehydrogenase*, an enzyme that breaks down and inactivates alcohol. Men's stomachs contain more of this enzyme than women's, which is one reason most women are more sensitive to the effects of alcohol than most men. Destruction of drugs by the stomach and intestines means that for an effective dose to reach the brain, more must be taken than if the same drug were inhaled or injected intravenously.

Drugs such as alcohol and aspirin must pass through the walls of the stomach and intestines and into capillaries that will deposit them into the veins. The intestines in particular are specialized to facilitate this process, but it still takes much more time for an ingested drug to reach the bloodstream than for a vaporized drug to pass from the lung to the blood. Even when the drug has reached the blood, it has one more hurdle to cross before it can reach the heart and then the brain. That hurdle is the liver.

All the blood that leaves the stomach and intestines passes through the liver before returning to the heart. This passage is important because the liver is loaded with enzymes that can inactivate (metabolize) all kinds of chemicals (including drugs) that might arise normally from metabolism or that we might ingest, either by accident or on purpose. In effect, the liver is a big filter for most chemicals that pass into the body from the gastrointestinal tract. So, after some of the ingested drug is lost in the stomach and intestines, even more is lost once the liver does its work. These two factors are the major reasons why medicines taken by mouth must be given in larger doses than those that are injected.

Although any substance that gets into the blood eventually passes through the liver, blood from the gastrointestinal system goes to the liver to be cleansed before it goes anywhere else. This cleansing of the blood on its first pass through the liver has a profound effect on reducing drug concentration in the blood. For example, the standard dose of morphine given in a hospital for severe pain is 10 mg when injected, or 60 mg when given by mouth. This sixfold difference reflects the effects of the gastrointestinal tract and liver on reducing the amount of drug that is delivered into the body.

But there is another difference between injected drugs and drugs taken by mouth. Because of the increased time required for absorption, the onset of the effects of a drug are slower when people take it by mouth than when they inject or smoke it. As a result, the peak effect of the drug may not be as striking. For example, addicts inject heroin intravenously to produce a rush. Heroin taken by mouth produces at best a small rush, even if enough is taken to produce blood levels equal to those attained through intravenous injection. Simply changing the route of administration from injection to ingestion eliminates one of the key effects sought by heroin addicts. We should not be surprised, then, that heroin addicts "mainline" (inject). Users of all kinds of drugs typically select routes of administration that maximize the drug effects they are seeking.

How Drugs Taken Through Mucous Membranes Reach the Brain

Mucous membranes, like the insides of the mouth and nose, are the linings of the various body cavities designed to allow things (such as food, air, or body wastes) to enter or leave the body. Unlike skin, they are kept moist, and certain kinds of chemicals can dissolve in this moisture and thereby gain entrance to the body. People can take a few drugs, like LSD (lysergic acid diethylamide), cocaine, and even heroin, into their bodies by absorbing them through the mucous membranes that line the mouth, nose, throat, vagina, and rectum. The moist surfaces that line the body's orifices allow passage of many kinds of chemicals and are near the blood supply.

Absorption through mucous membranes is not a particularly effective process, but it does work. People commonly take LSD by placing a "tab" under the tongue and letting the drug be absorbed there. (LSD is so potent that very small amounts, measured in micrograms, are effective. A dose of LSD is a tiny drop that is typically placed onto a small square of blotter paper. The drug user places the paper into his mouth, and his saliva absorbs the drug. The piece of blotter paper is the tab. Some of the other nicknames for LSD, such as "window pane" or "sunshine," refer to graphic designs made on the blotter paper.) Some medicines, notably the nitroglycerin that people take to treat angina, are also administered by this route. LSD abusers don't mind the slow onset of effects, because LSD does not produce a heroin-like rush by any route of administration.

In addition, people can snort cocaine or heroin into the nose, where it is absorbed from the upper respiratory tract. This route of administration is a little faster and more efficient than taking the drug by mouth and absorbing it through the stomach and intestines. Moreover, both cocaine and heroin leave a bitter taste that can be avoided by snorting them up the nose. If nothing else, snorting protects the drug from being digested. Because heroin is now so much more pure than it used to be, snorting it has become an effective and popular route of administration.

WHAT'S THE DIFFERENCE BETWEEN COCAINE AND CRACK?

When cocaine started to make its big comeback in this country in the 1970s, snorting was the major route of administration. At that time, people could not smoke cocaine (it was only available in powder form), because it burned rather than vaporized. Some users injected the drug, but most were not willing to do this. People could smoke cocaine by chemically converting it to its "free-base" form, but this was a dangerous process that required highly flammable chemicals and a flame to heat them with. To free-base was to risk being blown up or badly burned.

Then an underground chemist made a big breakthrough. He figured out how to make crack, which is free-base cocaine in solid form, commonly called a "rock." From then on, smoking cocaine was easy. All one had to do was buy crack from a dealer, heat the rock in a glass pipe or similar device, and inhale the vapor produced by heating it.

The introduction of crack ignited the cocaine epidemic that characterized the 1980s. Snorted cocaine had been producing a steady stream of addicts because cocaine is a powerfully reinforcing drug that quickly teaches users to keep taking it. But, even so, snorted cocaine is not nearly as reinforcing as smoked cocaine, simply because smoking can deliver a much larger amount of cocaine to the brain more rapidly than snorting. The rush that results from smoked cocaine is much more profound than that induced by snorting. The difference in drug effect produced by changing the route of administration is the pharmacokinetic reason why crack is so much more dangerous than powder cocaine. The drug is the same, so we can see what a profound difference changing the route of administration can make.

The absorption of drugs through mucous membranes generally is not particularly fast, and cocaine constricts blood vessels in the nose and decreases blood flow, limiting the amount of cocaine that can be absorbed over a given period of time. Ironically, cocaine has the opposite effect on the blood vessels of the lungs. Cocaine dilates these blood vessels and increases blood flow. So, crack cocaine not only has the huge surface area of the lungs through which to be absorbed, but by increasing blood flow through the lungs, the drug can be taken into the body very efficiently.

A sociological reason also contributed to crack's increasing cocaine addiction. Crack was not only a breakthrough in illicit chemistry, but a breakthrough in marketing as well. Before the introduction of crack, dealers sold cocaine in relatively large amounts for prices between $50 and $100 per gram (the standard unit of sale). A rock of crack sells for $5, $10, or $15. This has brought cocaine into the reach of young people — even poor young people — who are always at greater risk of addiction than adults. Because of its pharmacokinetic and marketing advantages, crack cocaine was a disaster for this country.

Injection Under the Skin

Some people inject heroin under the skin, a technique called "skin popping." This gives the drug more of a kick than snorting it, but not quite the punch that intravenous injection carries. "Skin poppers" are generally not addicts, but are drug abusers looking for a high while trying to avoid the perils of intravenous injection. However, skin popping is just a step away from intravenous injection, and people who use heroin this way are at high risk for addiction.

Transdermal Drug Delivery: Slow and Steady

The skin's job is to keep most things out of the body. It is very good at its job. However, certain chemicals can pass through the skin into the bloodstream. This is a particularly effective route for drugs that must be administered in steady doses over long periods of time. Nicotine, nitroglycerin, and some opiates all are drugs that can be delivered transdermally. This route is effective for medical use, but has no advantages for the abuser.

THE RATE OF DRUG DELIVERY IS IMPORTANT

In addition to the *route* of administration, the *rate* at which the drug reaches the brain is also important. For example, injecting heroin is popular among opiate addicts because it delivers heroin into the brain so swiftly. In fact, heroin is really nothing more than two morphine molecules linked together by a simple chemical reaction. Once heroin gets into the brain, the body breaks it down again into morphine, which then binds with the opiate receptor. Why then do addicts use heroin instead of morphine? Because heroin can pass through the walls of the brain's blood vessels and into the brain more rapidly than morphine can, allowing brain levels of morphine to rise more rapidly than if addicts injected morphine itself.

Slowing delivery to the brain does not alter all of the effects of a drug. The hours of analgesia and contentment and the degree of constipation that inevitably accompanies opiate use remain the same whether addicts inject heroin or take it by mouth. Changing the rate of drug entry to the brain alters the intensity of the initial high, the rush. Experienced drug abusers all recognize the importance of this. Drinkers, for example, may down several drinks in rapid succession because they know that if they can get a lot of alcohol into their blood quickly, the initial high will be more intense. So, the ultimate amount of drug in the blood has a crucial influence on the drug's effects, but the speed at which that amount reaches the brain will influence some of those effects.

THE TIMING OF DRUG DELIVERY IS IMPORTANT

A third factor also influences the "high" that drug abusers seek. Many drugs produce euphoria only when blood levels of the drug are rising for the first time during a bout of drug use. This is readily apparent with alcohol. As drinking starts and blood alcohol levels climb past a certain point, the drinker typically feels euphoric and has increased energy; he or she is high. But even if the amount of alcohol in the blood continues to climb, which it will do as long as the rate of drinking exceeds the rate at which the body destroys alcohol (about one drink per hour), the initial feelings of euphoria start to disappear after a while. Euphoria will not reappear as blood levels fall, even when they reach the same point at which the drinker first experienced it. That is, rising blood levels of alcohol produce effects that do not recur when blood levels fall. What's more, if the drinker stops for awhile and then starts up again and blood levels go back up, he will either experience no euphoria at all, or it will be greatly blunted. The period of euphoria and activation appears to be a one-time event for each bout of drinking.

The same phenomenon occurs with marijuana, cocaine, and many other drugs. The first episode of drug administration on a given day will produce a more intense high than later episodes. Nevertheless, this doesn't stop people from using drugs for extended periods, like smoking three or four joints during the course of a day, drinking all night long, or going on a binge of cocaine use that lasts for several days. But it does demonstrate that the simple binding of a drug to its receptor is not always sufficient to describe all the psychoactive effects of abused drugs. Something else happens the first time a drug is used that simply doesn't happen on subsequent occurrences. We don't yet know what it is.

HOW THE BODY ELIMINATES DRUGS

The body starts to eliminate drugs almost as soon as the drugs are absorbed. Drugs taken by mouth pass through the liver before going anywhere else. So, the body eliminates part of the alcohol in each drink before it can exert an effect. When people take drugs by other routes, these drugs bypass the liver on their first trip to the brain, but begin to show up in the liver's blood supply before too long. This is because all blood mixes in the heart and must traverse the liver before returning to the heart from the body. This is no accident. The liver is the body's most important site for inactivating a whole array of chemicals, including the normal byproducts of metabolism. It contains a number of enzymes that break down or alter all kinds of substances, including drugs.

Enzymes are large molecules that living organisms use to facilitate the transition from one form of a chemical to another. A chemist (or a baker) might mix together precise amounts of carefully chosen ingredients and heat them to bring about this transformation. The heat supplies the energy needed to change chemicals. But the body cannot use heat as a source of energy. Our cells have no Bunsen burners. Instead, we use enzymes, which can do the same thing, but without heat. In effect, enzymes are the workers on the intracellular production lines that make our bodies function. Enzymes help to build large molecules, like proteins, by assembling the amino acid building blocks together. Other enzymes break down larger molecules into smaller ones. A whole variety of enzymes takes part in the process of extracting energy from sugar, leaving only water and carbon dioxide as the end products.

From Active Drug to Inactive Metabolite

We commonly use the term *metabolism* to describe how the body's enzymes extract energy and other nutrients from food by breaking it down into components our cells can use. But metabolism has a broader meaning as well, and it includes all the processes by which the body breaks things down. Thus, the stomach and intestines contain metabolic enzymes that digest (break down) food, whereas cells contain other kinds of metabolic enzymes that extract the energy from sugar and fat. The liver and many other places in the body contain enzymes that break down or inactivate drugs in other ways. These enzymes are also called *metabolic enzymes*, and the products of their transformations are called *metabolites*.

In addition to the liver, the blood, brain, and other organs contain enzymes. Together, all these enzymes turn active drugs into *inactive metabolites*. The body inactivates alcohol by metabolizing it into smaller components. In contrast, it inactivates morphine by adding something to it.

The body has two goals in metabolizing drugs:

- To simply change the drug's shape so that it no longer can fit its receptors

- To create a chemical that will easily pass into the urine or feces so it can be excreted

Once the chemical structure is changed, the drug becomes inactive, because it will no longer fit its receptor. If the drug will not work at its receptor, it cannot exert its effects.

Metabolizing drugs not only inactivates them, but creates a resulting product, a different chemical, which is now easier to excrete than the drug it came from. Once the body has accomplished the task of metabolizing a drug, it can eliminate the new chemical through the urine, feces, or even the lungs. Some of the urine tests that determine whether a person has used drugs detect drug metabolites, not the parent drug, simply because the body is pretty efficient at eliminating the drugs themselves.

Some drugs, like nicotine, remain active for only a short period of time. The effects soon wear off. Other drugs, such as LSD, last for hours. The effects of methadone, a synthetic opiate used to treat cancer pain and heroin addiction, last for a day. If the body is well equipped to handle a certain drug, that is, if it has enzymes that efficiently transform the drug into an inactive metabolite, the drug will have only a short duration of action. On the other hand, if the body does a poor job of enzymatic inactivation, the drug will have a longer course of action. Thus, a drug's duration of action depends on the body's ability to inactivate it and then get rid of it.

To be able to work on a drug, an enzyme must have a structure the drug can fit into in much the same way that it fits into a receptor. Because drugs are not endogenous components of the body, our bodies are not equipped with enzymes designed specifically to deal with drugs. Instead, the enzymes that inactivate drugs were originally designed for compounds normally found in the body and can work on drugs only by coincidence.

STORAGE

Enzymatic inactivation is not the only process that determines the time course of drug action. For example, some drugs pass directly into the urine or feces without being altered. The rate at which this happens helps to determine how long such drugs will work. The body may store other drugs in its tissues, especially fat, which makes the drugs unavailable to the blood or brain. Some barbiturates go into the body fat; this shortens their effects because the body eliminates them from the blood just as if they had gone into the urine.

The body readily stores the THC from marijuana in its fat. This is one reason why THC disappears from the blood within a few hours after smoking. While blood levels are high, THC enters the fat readily, but it leaves *slowly* once blood levels have dropped. As a result, it is possible to detect both THC and its metabolites in urine for up to 1 week after a person has smoked a single joint. The amount of THC in the blood is far too small to produce the well-known psychoactive effects of marijuana, but sensitive chemical tests can still detect it in urine. In fact, a urine test for THC can determine whether a person has used marijuana, but it does not necessarily mean that the person has been intoxicated in the last few hours or even days. A blood test would be needed to determine the amount of THC in the blood and thus when the drug was taken.

Pharmacokinetic characteristics of different drugs, including route of administration and rate of elimination, can affect both peak effects of a drug and its duration of action. As demonstrated by the epidemic of cocaine addiction that resulted from the introduction of crack cocaine, these characteristics can have profound implications for drug abusers. However, the fundamental pharmacological properties of drugs depend on the kinds of receptors they interact with and the cascade of events they initiate, rather than how the drugs got there.

CHAPTER 7

DRUGS TELL THE BRAIN TO TAKE MORE DRUGS

Barry

Barry's mother, Cynthia, pulled into the garage and shut off the ignition. She was completely bewildered. What was going on? They were a typical American suburban family. Her husband Richard was a neurologist; she was a volunteer who raised money for the symphony and the art museum. Their children, Barry, 17, and Lisa, 14, were the center of their lives. They were wonderful kids, bright, good students, active in Little League, soccer, scouts, and a dozen other activities as they were growing up, now completely absorbed with the world of high school and friends — perfect teenagers.

Well, not *perfect*, a nagging voice intruded. Barry's grades fell last semester. But that was because he took difficult courses, and one of his teachers, Mr. Monroe, was a real jerk, Cynthia argued with herself. So far as Mr. Monroe was concerned, Barry couldn't do anything right, poor kid.

But I don't see much of Barry anymore, the voice continued. Well, true, he's become somewhat withdrawn this summer, Cynthia mused, less involved with the family and more engaged with his friends, staying out late at night and sleeping through most of the day. But that's adolescence . . . isn't it? She gathered the groceries and her purse and went inside to start dinner, still preoccupied with the news the bank had given her an hour ago.

Cynthia and Lisa were just sitting down to eat, when Richard arrived, late from rounds. The family had long since learned to share Richard with his patients. Cynthia fixed him a plate and handed it to him as he entered the dining room and joined them.

"The strangest thing happened this afternoon," she blurted out.

Cynthia hadn't meant to tell Richard until later, after he'd finished his dinner, but she was so upset she couldn't stop herself. "John Williams from the bank called. Someone made 22 ATM withdrawals from our accounts this month."

"What?" said Richard.

"$500 each withdrawal," she replied. "$11,000 gone, just like that. And my card is still in my wallet. You haven't lost yours, have you?"

"Let me look," Richard said, checking his wallet. "No, it's right here. Damn it, Cynthia, why do you always hit me with this stuff just as I walk in from a tough day?" They quickly moved into yet another argument, which was becoming a nightly ritual.

"Stop it, STOP IT," Lisa shouted, stunning them both into silence and starting to cry. "It's Barry, Mom. He sneaks the card out of your wallet, withdraws the money, and puts the card back before you miss it."

"What?" said Cynthia and Richard simultaneously, both dumbfounded. "Why would our own son steal from us?" asked Cynthia.

"You don't get it, do you Mom?" Lisa gasped, pushing out the painful words between sobs. "Mr. Monroe's been trying to tell you. Barry's on drugs. He's in big trouble. And I'm scared."

Cynthia's question is a good one and brings us to the essence of addiction, the state in which drugs control a person's behavior so he or she can no longer control it. How could someone you love and trust unconditionally — your own child — steal from you? How could that child's behavior change so imperceptibly that his own mother and father can't see it. Yet, so drastically, it baffles them when they do? How could anyone do this to himself and to the people who love him most?

The answer to all of these questions is that Barry has become a drug addict. Addiction is the loss of control of drug-taking behavior. The addict cannot stop using drugs despite adverse social, legal, and health consequences. Barry has lost control of his drug-taking behavior. Just about everything he does is driven not by his commitments to his schoolwork, friends, or family, or even by his own free will, but by his overwhelming need for drugs. He is preoccupied with getting drugs and will do almost anything to get them and keep taking them, even if it means missing school, lying to people about where he is and what he is doing, or stealing money from his parents. How did Barry, a young man who insisted on making all of his own decisions, get into a situation in which some *thing* is making his decisions for him? It took the actions of the drug on his brain. It took time. And, it took lots of learning.

HOW ADDICTION EVOLVES

Addiction occurs gradually, but with fairly predictable milestones marking the way. With most drugs, each level of involvement — (1) experimentation, (2) more regular use, (3) tolerance, (4) physical dependence, (5) psychological dependence, and (6) addiction — develops over time, which is why Barry's parents couldn't see it. The people closest to an addict often don't see it.

As each stage of addiction progresses to the next, the brain changes, and each of these changes alters behavior. At first the changes are subtle, and easily reversible. But they become increasingly overt and more difficult to undo as the addictive process progresses. Barry kept giving drugs the opportunity to change his brain, so drugs gradually began to dominate his decision-making ability. In part, they did this by altering brain functions and in part by using brain circuits that underlie normal behavior, especially circuits that we normally use to learn.

LEARNING

Much of what happens along the road from drug experimentation to drug addiction is learning, and with each stage of the addictive process a great deal of learning takes place. To understand what this means, let's look at how we learn and at how the knowledge that we acquire shapes our behavior.

Nearly every form of human behavior involves some kind of learning. Learning is the way we gather knowledge about our world. Once we have stored that knowledge, it is *memory*. Specific brain processes carried out in identifiable brain circuits allow us to take in information from our environment, store that information in memory, and retrieve it when we need it. What we learn from our environment plays a crucial role in how we behave.

Let's say you see (that is, take in information about) a man walking towards you on a busy city street. If he has a smile on his face, is walking casually, and passes by others in a friendly manner, you will walk by him without much thought and go about your business. However, your behavior would change radically if this person scowled, swaggered in an aggressive manner, and frightened others away as he approached you. Sensing potential danger, you would become more vigilant. You might keep walking toward him carefully; or, you might cross the street, duck into a store, or take some other defensive measure to avoid this potential threat.

As children, we often do not recognize the threat posed by some people in some situations like that, but in the course of our lives, we learn how to read environmental cues to protect ourselves. First, we have learned how to read hostile, aggressive body language. Second, we have learned to do something about it when a person displaying these behaviors approaches. As this example illustrates, what we learn from our environment influences how we behave.

EXPLICIT LEARNING AND MEMORY

Learning occurs in two fundamental forms: conscious and unconscious. With *conscious learning*, which psychologists also call *explicit learning*, we use our senses to find out what is in the world and where and when events have occurred. We take in information about what we see, hear, smell, taste, and touch, and our brain stores that information in its short-term, or *working*, memory. Our brain's working memory allows us to retain information long enough to perceive it, use it, perhaps mull it over a bit, and then either store or discard it.

For example, as you enter the supermarket to buy the things you need for tonight's big dinner, you walk through the aisles to see what else looks appealing. As you head for the tomatoes, you keep a mental note of where these things are located. After picking out the tomatoes, you head back to pick up the fresh-looking lettuce and the crisp green beans you saw earlier. You selected these things because you had noticed they looked fresh, stored that observation in your short-term memory, and got back to it once you had completed your most important task. After you leave the store, however, you almost certainly will forget which vegetables looked good or bad, because you need your working memory to do other things, and the information about those vegetables is just not important enough to store in long-term memory.

If you use a computer, you've probably written a short note or memo, e-mailed or printed it, and then closed the document without saving it. You kept the information on the screen (and in the computer's "short-term," or buffer memory) long enough to use (send or print) it. Then, when the information was no longer useful, you discarded it, much like your short-term memory does. On the other hand, after you've entered an important document on the screen, you save it to the hard drive so you can retrieve it later. Just as your computer transfers your document from its buffer memory to its hard drive, your brain can transfer important information from your short-term memory to your long-term memory.

Our *long-term memory* is like information stored on a hard drive. We can recall at will information that we use frequently, like people's names, or information that we have not used for a long time, like our street address from childhood. We use some things frequently, like the names of our children, and we have therefore learned them well. Studying to pass a test is the process of using some information just frequently enough to store it. Other times, we store memories because they are accompanied by a strong emotion. For example, almost everyone who is old enough remembers where he or she was when the Challenger exploded and when President Kennedy and Martin Luther King were killed.

In addition, we remember important personal events that aroused particularly strong emotions and were especially meaningful to us, though perhaps not to most other people — our wedding, the birth of a child, the death of a family member, and similar milestones.

Like most drug users, Barry remembers the first time he tried marijuana. A friend gave him some potent pot, and the effect it had on him was so intense that he will remember it for a long time. Each new drug he tried over the next year or so — methamphetamine, MDMA, heroin, and others — had similar intense effects the first few times he used them. That powerful memory of early drug experiences is one of the things that drives users to keep taking drugs. They keep trying to duplicate the memory of those first few highs.

Short-term versus Long-term Memory

There are crucial differences between short-term memory and long-term memory with respect to the brain. With *short-term memory*, information coming in from the outside world circulates along a complex brain pathway, which begins at the sensory cortex and passes through the hippocampus. The hippocampus sends the information on to the prefrontal cortex. Neuroscientists believe that rather than stopping in the cortex, incoming information flows continuously along this circuit until it is discarded.

With *long-term memory*, instead of briefly circulating the information, the brain fixes the information in the cerebral cortex. It now appears that long-term memories are laid down in the same parts of the cortex where they are first perceived. Thus, the memory of the sight of an apple would be stored in the visual cortex, but the memory of its taste would be stored in the part of the cortex where we process information from our taste buds. Understanding how the brain stores memories is one of the great quests of neuroscience. There is now considerable agreement about where in the brain this takes place (at least for vision), but precisely how the process works is still mostly a mystery.

Most neuroscientists believe that the neurons in the long-term memory circuit probably undergo some form of structural change almost certainly at their synapses. Under the proper circumstances, neurons may generate new dendrites, axon terminals, receptors, and other components of the synapse to strengthen their connections to one another and thus increase the effectiveness of synaptic transmission. The best bet right now is that increased effectiveness of synaptic transmission underlies long-term memory.

Disruption or Destruction of Memory

It is important to understand where the brain stores memories for a number of reasons having to do with drug use. Marijuana disrupts the hippocampus and therefore disrupts short-term memory. The disruption appears to end as the drug is eliminated from the blood, but while the person is intoxicated, and probably for a few hours thereafter, the drug impairs short-term memory. If information fails to remain in short-term memory, the brain cannot reliably transfer it to long-term memory. This is one reason Barry's grades fell last semester. Because he used drugs so frequently, he often tried to study when he was stoned, but his brain could not transfer to his long-term memory the information he was taking in.

Severe alcoholism, in which the drinker consumes large amounts of alcohol for many years, can produce a brain disorder called *Wernicke-Korsakoff's syndrome*. This disease damages some of the brain circuitry needed to create new memories and so disrupts the process that puts new information into long-term memory. This impairment is not limited to periods of intoxication. In extreme cases, people with this disorder can never learn anything new again.

Patients with Wernicke-Korsakoff's syndrome have been studied for many years. One such patient, a man in his 50s, had been addicted to alcohol all of his adult life. You could engage him in a fairly long conversation, and, after being introduced, he would repeatedly refer to you by name. He could also recall all sorts of memories from childhood and early adulthood, but nothing from the later part of his adult life after alcohol had destroyed his ability to store new information. If you were to leave the room during a conversation with him and then return a few minutes later, this patient would stare blankly at you and ask, "Who are you?" Because he no longer had the ability to store new memories, he could not remember ever seeing you. He could keep information in *working memory* while he was using it, but things just disappeared from his awareness forever not long after they were out of sight or out of the present.

Learning Can Help or Hurt

Basically, structural changes in the synapse that enable people to learn and retain what they've learned can be good or bad. People who proceed down a fairly normal path of development throughout childhood and adolescence can learn many behaviors that help them get the things that they want — a good job, meaningful relationships, and the ability to enjoy and appreciate the subtle pleasures of life. They can live out their lives in positive ways and even help better their world. But people also have the capacity to learn *maladaptive behaviors*. These behaviors are as difficult to unlearn as the adaptive behaviors would be, but the need to unlearn them is one reason addiction is such a difficult disorder to treat.

IMPLICIT LEARNING AND MEMORY

Drug users learn lots of maladaptive behaviors. They acquire some of them through the kind of learning we have just described, that is, through conscious, explicit learning. But they also acquire them through a second kind of learning, *unconscious* or *implicit learning*. Implicit learning shapes our behavior, in some ways much more profoundly than conscious learning, but we are not aware that we are learning while the process is taking place. One kind of implicit learning teaches us how to do things (as opposed to what they are or when they happened). Other kinds of learning teach us to associate a stimulus either with another stimulus or with a behavior. These kinds of learning take place through repetition. Although they take practice, that's all they take.

OPERANT CONDITIONING

Barry at first made the decision to use marijuana and other drugs of his own free will. But the very first time he used them, the drugs he took initiated a powerful form of unconscious learning called *operant conditioning*. Over time, operant conditioning produced profound effects on Barry's behavior and reinforced those effects each time he used drugs. It is the same with all drug users.

Operant conditioning teaches us to link a stimulus and a behavior. Although we are not aware that we are learning this relationship, we learn it all the same. Operant conditioning's teaching method is reward, and its best teaching tool is good feelings. It reinforces what we do by making us feel pleasure. For example:

- We do something (behavior).

- It makes us feel good (reward).

- We do it again (repeat behavior).

The reward reinforced the behavior. The more the behavior is reinforced, the more likely we are to repeat it. Thus, we gradually learn to repeat behaviors that make us feel good.

Positive and Negative Reinforcement

Natural reinforcers, like food and sex, are *primary reinforcers*. They directly produce positive reinforcement, with no learning or intervening steps required. Drugs do this as well. *Secondary reinforcers* produce reward, but not as directly. Their value as a reinforcer has to be learned. Money is a secondary reinforcer. Valueless by itself, you can't eat or inject it, money has immense power to motivate people. For some, being thin is a secondary reinforcer; for others, it is having a muscular body. These secondary reinforcing properties can also support drug use. For example, smoking cigarettes helps keep people thin. If they give up smoking, they will gain weight, an undesirable outcome. Men especially, but women too, abuse anabolic steroids because the bulked-up muscular look these drugs help to produce is a secondary reinforcer. It may be that people abuse hallucinogens because of their secondary reinforcing properties. These drugs relieve boredom and enhance emotionality. These effects are not primary reinforcers, but desirable for some people nonetheless.

The reward produced by primary reinforcers is called *positive reinforcement* because it produces feelings of pleasure. Operant conditioning can also take place if bad feelings like anxiety, stress, or pain are removed. When this happens, the reward is called *negative reinforcement*.

- We feel bad.

- We do something (behavior).

- It alleviates the bad feelings (negative reinforcement).

- The next time we feel bad, we do it again.

Negative reinforcement increases the likelihood that the behavior that relieved the bad feelings will be repeated. The learning process is the same whether the reward is positive reinforcement or negative reinforcement. Positive and negative reinforcement are two sides of the same operant conditioning coin.

We can see how positive and negative reinforcement works when we think about very young children. When a child gets tired or frustrated, he often starts to cry (behavior). His mother or father picks him up and comforts him, relieving his distress and making him feel better (reinforcement). The next time the child feels distress, he cries again (repeats the behavior). He's learning a relationship between crying and getting attention, and the relief he gets when mom or dad pick him up is the tool teaching him this lesson.

Similar lesson plans teach babies how to get fed when they are hungry, and, when they get much older, teach them to have sex in order to perpetuate the species. Think about the first time you kissed someone and felt the stirring of sexual desire.

- You kissed someone (behavior).

- It made you feel wonderful (reward).

- You kissed again (repeat the behavior).

As you explored the additional steps that ultimately lead to having sex, the good feelings they produced led you to repeat these newfound behaviors as often as possible. You do something. It makes you feel good. You do it again.

Where Operant Conditioning Occurs: The Brain Reward System

Through operant conditioning, our brain teaches us to do things that help us get food to eat, liquids to drink, sex to create the next generation, and all the other things we must do to ensure our own survival and that of our species. Operant conditioning is so important that it is linked to a special neural circuit within the brain's limbic system. When we do something to activate this circuit, it teaches us to repeat that behavior by releasing a cascade of neurotransmitters that suffuse us with feelings of pleasure. Because it does this, many people call it the *pleasure circuit*. Scientists call it the *reward circuit*, or *reward system*. Anything that turns this system on reinforces the very behavior that turned it on in the first place. We learn to do things that turn on our brain reward system.

Although the kind of learning that operant conditioning teaches us takes place all the time, we are almost always unaware of it. But we can see how it works in the laboratory when we teach a rat to press a bar to obtain food. Bar pressing is not something rats living in the wild typically do. But once a rat accidentally presses the bar in a training situation and obtains a food reward, its interest in the bar increases. The rat is not quite aware of what it did, but it understands that being near the bar produced food. Simply by spending more time near the bar, it is more likely to press it again. A few accidental presses that produce food reinforce the rat's interest in the bar again, until it learns that pressing the bar is the behavior that produces a food pellet. Then, a hungry rat will spend lots of time working on that bar.

The same is true for people. We often use the technique of operant conditioning to get people to do what we want them to do. If they do something for us, we give them a reward. It may be as simple as a "thank you" or as powerful as money. We teach children to behave well by giving them praise and sometimes special treats when they exhibit good behavior.

UNDERSTANDING BEHAVIOR

The Discovery of Operant Conditioning

Edward Thorndike and B.F. Skinner discovered operant conditioning in their work with animals. In fact, Skinner developed a special chamber for studying behavior. This chamber, which came to be known as the "Skinner Box," was the first device that automatically released food and measured behavior over long periods of time. Skinner showed that instead of merely observing behavior, it could actually be measured objectively and altered in predictable ways by manipulating the reward.

Accidentally stumbling on behaviors that turn on your brain's reward system takes time. Then, you have to repeat these behaviors to learn their lessons. Like the rat who gradually learns to press the bar to get a food reward, you try out a lot of behaviors and gradually learn that some of them turn on your brain's reward system and make you feel good. Sometimes, because human interactions are so complex, it's hard to pull out the behavior that is producing reward. What Skinner did was to supply a technique that could identify the key behaviors in a situation that might otherwise be too complex to study effectively.

Skinner's work profoundly influenced so many researchers that a whole generation of psychologists used his technique to study animal and human behavior. These researchers were so focused on studying behavior, however, that they purposefully disregarded the brain. It fell to neuroscientists to combine the study of behavior with the study of the brain.

THE BIOLOGICAL BASIS OF BEHAVIOR

Mapping out the precise sites in the brain that constitute the reward system has taken many years, and all the pieces of the puzzle are still not in place. Based on what they have discovered about the brain thus far, neuroscientists are teaching us that all behaviors and psychological processes have a *neural substrate*. These substrates consist of interconnected physical locations in the brain with definable patterns of activity that generate the behavior or psychological state. In short, a biological basis for behavior.

Drugs Teach People to Take More Drugs

The absolutely stunning thing about drugs of abuse is this: they turn on the reward system directly. No trial-and-error effort is required. Unlike all the natural rewards, such as food and sex, which activate this system subtly, slowly, and indirectly by first activating other brain structures, drugs activate the reward system directly. Positive reinforcement can be rapid and powerful: if natural reinforcers turn on a light, drugs set off fireworks. Drugs have the power to do this because they mimic or alter the actions of the brain's neurotransmitters in our reward system. And, instead of turning the reward system on subtly like a natural reinforcer would, drugs flood in and do it with a jolt. The direct effect of the drug on this system causes reward. Thus, through operant conditioning, *drugs reinforce drug-taking*. Nothing else has to happen. By activating this neural circuit, instead of teaching people to pursue survival, drugs threaten their survival *by teaching them to take more drugs*.

Drugs Teach Animals to Take More Drugs

Laboratory experiments with monkeys illustrate just how powerfully drugs turn on the reward system. Scientists insert a special catheter into a monkey's vein. The catheter is hooked to a pump that pushes a specific dose of cocaine into the blood each time the monkey presses a bar to which the pump is attached. This monkey has been trained to press bars that release food and water, but now the scientist has introduced a third bar, and pressing this one releases cocaine into the monkey's bloodstream. If the monkey is given unlimited access to cocaine, it will press the cocaine bar over and over again. It will ignore food. It will ignore water. It will press that cocaine bar endlessly to get more and more of the drug. In fact, if the scientist doesn't stop the experiment, the monkey will keep taking cocaine until it dies.

Drugs Taught Barry to Take More Drugs

The very first time Barry used marijuana and other drugs, he just loved the way they made him feel. We now know why. Each drug he used acted directly on his brain's reward system. Because drugs can do this, each drug made him feel intense pleasure. Barry's behavior then followed the rules of operant conditioning: He did something (took drugs). It made him feel good. He did it again (took more drugs). Drugs taught Barry to repeat drug use and reinforced that lesson each time he did.

CLASSICAL CONDITIONING

Another form of implicit learning, *classical conditioning*, builds on the foundation of operant conditioning. Classical conditioning teaches drug users to associate ordinary things and events with drug use. Over time, these ordinary things can provoke craving, intensifying the motivation to take more drugs.

Discovery

Ivan Pavlov discovered classical conditioning in his work with animals. While feeding the dogs he was studying, Pavlov noticed that the animals began to salivate long before he placed food on their tongues. Hypothesizing that something besides food must be triggering this response, Pavlov set up an experiment to test his idea. Every time Pavlov got ready to feed his dogs, he first rang a bell. Soon the animals learned that the ringing bell signaled they were about to be fed, and they began to salivate at the sound of the bell. The animals became conditioned to the bell's ringing. They learned to associate a previously meaningless stimulus — a bell — with one that was important for their survival — food, which they needed to eat. The simple repetition of the paired presentation of the ringing bell with food taught the dogs that the ringing bell meant food was coming. Eventually, the bell produced the same response as food: salivation; thus, classical conditioning.

Cues

Classical conditioning teaches drug users to associate ordinary things such as talking on the telephone with the act of drug-taking. For example, when Sybil first started smoking, she often lit up a cigarette while talking on the phone. Gradually, Sybil unconsciously learned to associate talking on the phone with smoking a cigarette. After a while, the association became so strong that every time she answered or made a phone call, Sybil wanted a cigarette. For Sybil, a phone call became a stimulus, more commonly called a *cue* or *trigger*, which actually provoked craving for nicotine. Without ever realizing what had happened, she satisfied that craving by lighting up.

Drug users develop dozens of cues that work to motivate them to keep taking drugs. Many cocaine users roll up a $10, $20, or $100 bill and snort the drug into their nostrils through this expensive tube. They learn to associate a neutral piece of paper — a $10 bill — with the act of snorting cocaine and the feelings the drug evokes. After a while, the sight of a rolled-up bill becomes a cue that provokes a craving for cocaine in these drug users. In fact, because drug addicts constantly need money to buy the drugs to which they are addicted, the sight of any money can provoke craving. Henry's "works" (the paraphernalia he used to inject heroin) became a trigger for him. So did the mere sight of his drug dealer, Jake, and even the apartment out of which Jake worked. And smelling the glistening white powder also became a powerful trigger for Henry, who could not then restrain himself from shooting up. Because of classical conditioning, triggers such as smelling the drug, seeing his works, and seeing Jake or Jake's apartment, all of which had once been neutral stimuli, eventually gained the power to trigger in Henry powerful cravings for heroin.

Barry, too, has developed a whole array of triggers, among them his drug-using friends, his bedroom, and his car, where he uses drugs without letting his family know. Even the ATM machine where he withdraws money from his parents' checking account has become a cue because he goes right from the machine to his dealer.

Drug users develop triggers as soon as they start using drugs. The double-wide papers Neil uses to roll joints are becoming a trigger for him. Each time he sees them for sale somewhere, he begins to think about marijuana. Michelle thinks about the drug each time she sees Neil. Polystyrene coolers make Allison think about beer. At first, these triggers are not very strong, but they acquire increasing power as they are repeatedly associated with drug use over time.

Thus, through classical conditioning, drug users unconsciously learn to associate drug-taking with many ordinary, everyday things. Classical conditioning does not cause addiction. It is a form of learning that occurs while people are on the road to addiction. The triggers in developing addicts are not particularly strong, and they use drugs whether or not their secret triggers are activated. As time goes on, triggers add power to the urge to use drugs, but while drug use continues, the more powerful forces of reward and withdrawal control the addict's behavior.

RELAPSE

Although classical conditioning is not responsible for addiction, it would be a mistake to underestimate its role in relapse. It is precisely during this period, when addicts are trying to avoid drug use, that unanticipated cravings can undermine good intentions. During recovery, addicts are fighting well-entrenched behaviors that lead to drug use. Anything that causes them to think of drugs or that elicits craving makes that fight harder. Many of the better treatment programs help addicts deal with triggers.

ADDICTION: THE SUM OF MANY KINDS OF LEARNING

We now know that repeated drug use changes the brain. Addictive drugs that people use exert their effects through their direct actions on axon terminals, receptors, and other parts of the synapse. And learning reinforces these changes and even gives them their meaning (implications for behavior). This learning uses the same neural circuits and processes as the learning we acquire under normal circumstances. Like any lesson well learned, it remains long after the teacher has left the classroom and the student has graduated.

Addicts, who have advanced degrees in drug use, have retained explicitly learned memories of powerful early drug experiences. They have learned about the rewarding effects of drugs. By repeatedly practicing this lesson, they have powerfully reinforced it. Moreover, addicts have acquired an impressive repertoire of unconscious, classically conditioned responses that will make them crave drugs when exposed to certain environmental stimuli — stimuli that have no special meaning for nonaddicts. In effect, the learning processes that we must undertake to survive have been commissioned to help addicts learn behaviors to get drugs into the body on a regular basis, no matter what obstacles might be in the way. These behaviors are maladaptive for anyone who wants to live a normal life.

The learning of maladaptive behaviors is another key aspect of addiction. It is a difficult task to learn to be a functional adult with flexible responses to complex and painful personal and social situations. The prolonged human childhood created by evolution and the extended period of formal education and training that society has provided for adolescents are clear responses to the degree of this difficulty. In young people, drug use interrupts and takes over this maturation process.

When young people use drugs to solve problems, they forfeit the opportunity to develop and practice adaptive responses. How does an adolescent learn to deal appropriately with the pressures to succeed in school, with the disappointment of being dumped by a boyfriend or even with the complex experience of falling in love? If she "gets high" to deal with or avoid these experiences, then she never learns responses that can build the foundation for a happy and productive life. When adults use drugs to cope with their problems, they, too, undermine the adaptive responses they might have learned, and they fail to develop the new ones that are needed to survive in our continuously changing world.

Therefore, another result of the learning that takes place during drug use is that other kinds of learning, the kinds we need throughout our lives, fail to take place or are undermined. It is true that the high school basketball star who spends all his time practicing spin moves and jump shots rather than studying is at risk of failing in college because he never learned how to study. It is also true that the high school student who abuses drugs is at risk of failing in life and becoming an addict because he never learned how to live. On top of all the other things drugs may have done, getting and using drugs just took too much time, so other activities got pushed aside.

As we explore the physiological changes that are part of addiction, such as tolerance and physical dependence, it is important to keep in mind the crucial role of *learning* in the creation of an addict. If a person repeatedly uses drugs as Barry did, physiological changes occur gradually, but inevitably. The learning we have described in this chapter — conscious, explicit learning and unconscious, implicit learning, including both operant and classical conditioning — occurs in parallel. At some as yet unidentifiable point in this progression — experimentation, more regular use, tolerance, physical dependence, and psychological dependence — a controllable *desire* for drugs turns into a compelling *need* for drugs. Scientists believe that this switch takes place because of a change in the brain's reward system. Once it does occur, drug users like Barry become fully addicted and lose control of their drug-taking behavior.

CHAPTER 8

How Drugs Change the Brain to Produce Intoxication, Tolerance, Sensitization, Physical Dependence, and Withdrawal

*P*rolonged drug use makes an array of changes in the brain. Tolerance, sensitization, physical dependence, withdrawal, psychological dependence, craving, and addiction all describe altered functioning in the brains and bodies of long-term drug users. These terms, which have explicit meanings for scientists, do not always retain such precise scientific meanings when people use them in general conversation. It is therefore sometimes difficult to discuss drug effects because we don't always know what people mean when they use these important terms.

To scientists, each term describes a different change in the brain. But there is not yet agreement about whether any or all of these terms describe changes in the brain that (1) disappear soon after drug use stops, (2) outlast drug use for a limited period, or (3) are permanent. If drugs produce permanent, or at least very long-lasting, changes in brain function, it would be hard to deny that drug addiction is a brain disorder. If, on the other hand, drug-induced changes reverse soon after drug use stops, then it would be more difficult to support this concept.

Normal Changes in the Brain

Before we examine drug-induced changes in the brain such as tolerance and dependence, let's look at some other kinds of changes in order to develop a broader understanding of what a change in the brain really means. This will allow us to have a more meaningful understanding of drug-induced changes and the role each change plays in abuse and addiction.

The brain is changing all the time, a property neuroscientists call *plasticity*. Many of these changes take place at the cellular or molecular levels, so they are not readily apparent. Even though we may not be able to see what's happening, many things can cause changes in the brain. One of the remarkable strengths of the brain is that it can adapt to changes in the environment around it.

The brain is a very plastic (changeable) organ. For example, its plasticity allows it to undergo the normal, healthy changes that underlie crucial functions such as the laying down of memories (learning). If our brains were not able to change, we could not learn, because learning almost certainly is the result of some structural change within synapses. It is not yet clear what this change is, but it allows certain groups of neurons to communicate more efficiently with each other. When the neurons of these groups are active together, we recall our memories.

ABNORMAL CHANGES IN THE BRAIN

Not all changes in the brain are normal or healthy. In some brain disorders, such as Parkinson's disease and Alzheimer's disease, neurons actually die. External injuries to the brain, such as blows to the head, can also kill neurons. So can internal injuries. Strokes, which occur when a blood vessel in the brain gets blocked, prevent blood from reaching part of the brain. Without blood, neurons become starved for oxygen and, like any other cell starved for oxygen, they die. Unfortunately, the brain has at best a limited ability to make new neurons, and right now it appears that once most neurons die, they are gone for good. Depending on where and how large the damaged areas are, the functions they participate in are either lost or diminished.

Some drugs, such as alcohol, methamphetamine, and MDMA (Ecstasy), can also kill neurons. After long-term exposure, alcohol can kill neurons in the part of the brain that helps create new memories. A person who is seriously affected simply cannot learn anything new. Methamphetamine has been shown to kill dopamine-containing neurons in animals, and MDMA can kill neurons that contain another neurotransmitter called serotonin. The animal data are pretty clear about this, and recent data from human MDMA abusers indicate that human users are at risk as well.

SURVIVAL OF THE FITTEST

Disease, injury, and abuse of some drugs are not the only reasons for the loss of nerve cells. Our brain gets rid of many of its own neurons when we are very young children. We are all born with many more neurons than we need. As we pass through infancy and early childhood, the neurons that survive are those that make working synaptic connections with other neurons. The neurons that fail to hook up die. This whole process resembles the approach gardeners use when growing vegetables from seeds. First, gardeners overseed the vegetable patch. After shoots emerge, they encourage the thriving seedlings by pulling out the shoots that are growing more slowly. In the brain, of course, there is no gardener. Instead, nerve cells compete with each other, and only the most successful survive.

Neurons Can Add or Subtract Parts

Not all brain changes require the loss of neurons. Other, less overt, changes can occur as well. Some changes are in response to the work that the brain has to do. For example, frequent use of some parts of the brain may cause more dendrites or axon terminals to grow. Extra dendrites or terminals enable neurons to receive or send messages more effectively.

Young rats raised in rich environments that contain lots of things to explore and play with actually have more dendrites on some of their neurons than rats raised in the boring environment of an empty cage. The brains of the rats in the rich environment are busy at work, and this induces the growth of dendrites. The same is probably true for humans. Babies whose parents provide lots of attention and a rich environment with lots of fascinating toys to explore (i.e., keeping the babies' brains busily working) have a better chance of reaching their full intellectual potential than babies who don't get these advantages. That's because stimulating a growing baby with lots of things to play with and with lots of attention actually changes the structure of the baby's brain.

CHANGES AT THE MOLECULAR LEVEL

Scientists can see and measure changes in dendrites and axon terminals simply by looking under a microscope. But scientists also can see and measure even smaller, more subtle changes, which take place at the molecular level. These changes include alterations to certain receptors, components of second-messenger systems (see Chapter 4), and even alterations in some of the molecules that help give the neuron its shape. Even though the number of neurons, dendrites, and axon terminals remains constant, neurons can increase or decrease the numbers of the smaller, molecular parts that make up these larger structures.

For example, a part of the brain reward system in cocaine addicts appears to contain fewer dopamine receptors than the number found in normal brains. (We will discuss why this may happen later in the chapter.) After exposure to opiates such as morphine, neurons that contain opiate receptors show an even more subtle change. Somehow, exposure to opiates makes the opiate receptors less effective in activating their second-messenger system. The number of receptors doesn't change, and the number of second messengers doesn't change, but the two are just not able to link up as well. The weakening of the linking process may be due to a change in the molecular structure of the receptor, the part of the second-messenger system it links to, or to another molecule that helps the linking process. Whatever the reason, morphine is less effective in getting the cell to respond to its message.

CHANGES IN GENES

More recently, scientists have begun to examine how changes in our environment affect the way our genes work. *Genes* are the strands of deoxyribonucleic acid (DNA) that contain the blueprints for all the molecules the cell makes. This blueprint is copied to a related molecule called *messenger ribonucleic acid* (mRNA), which carries it to a part of the neuron that uses the code to make proteins and other components of the cell. For example, if the cell needs to make more dopamine transporters, it might make more copies of the mRNA that carries the code for the dopamine transporter. Scientists can now measure the changes in the amount of mRNA, opening a window on the most fundamental level of control of cellular activity. Chronic use of cocaine appears to change the amount of mRNA that makes the dopamine transporter. And the number of transporters also changes.

BEHAVIOR CHANGES THAT REFLECT BRAIN CHANGES

Finally, scientists can measure some changes in the way the brain works even when they cannot show precisely where in the brain those changes take place. Historically, behavior has provided one of the most important ways to study brain function. When people behave in ways that are obviously different from normal behavior, it is reasonable to explore whether the brain has been changed.

People with depression, epilepsy, and Parkinson's disease all show changes in behavior. These changes are so obvious that the behavior itself is diagnostic of the disease. In fact, doctors used behavior to diagnose these diseases long before they knew what brain changes cause these diseases. Now, because of everything scientists have learned over the last century, especially in the last few decades, we are able to identify specific structures in the brain that various diseases and disorders damage or alter.

Linking Parts of the Brain to Specific Functions

To understand what was going wrong in disease, scientists first had to understand which parts of the brain produced specific functions. The first step, begun in the last century, was to correlate specific behaviors with identifiable brain regions. Two neurologists, a Frenchman named Broca and a German named Wernicke, had dramatic success with this approach. They discovered the parts of the brain that allow us to speak and understand language, and they did this by studying people who had lost their language abilities.

Broca had patients who could understand language, but couldn't speak or write intelligibly. These patients simply couldn't generate the motor commands needed to create language. Wernicke had patients who couldn't understand language, but could speak (although what they said didn't make any sense). After their patients died, Broca and Wernicke studied their patients' damaged brains. They were able to relate their patients' lost language abilities to damage in specific parts of their brains.

The part of the brain that allows us to recognize words is located where the occipital, temporal, and parietal lobes all meet, now called *Wernicke's area*. The part of the brain that allows us to speak and write is a region of the frontal cortex, called *Broca's area*, which is part of the motor system.

CHANGES IN BRAIN METABOLISM

Scientists still try to identify regions of the brain that support different functions by studying people with various behavioral problems. Recently, powerful brain imaging techniques have become available to help scientists do this. These new techniques allow us to watch the brain in action. One of these techniques is called *positron emission tomography* (PET) scanning.

In PET scanning, a person lies inside a circle of special sensors that are like very sensitive Geiger counters. The scanners detect the location and amount of special radioactive chemicals that are injected to measure brain function. Computers accumulate the information from all the sensors and determine the amount of radiation coming from different parts of the brain.

Glucose and Energy in Your Brain

Because scientists already know that the brain uses glucose, a simple sugar, as its major source of energy, they reasoned that they could identify active brain regions by measuring how much glucose those regions used. To do this, they placed a radioactive tag onto a modified form of glucose and injected it into volunteers who were inside a PET scanner. Experiments based on this approach show that when a specific part of the brain engages in a task, it uses more energy than other brain regions that are not so busy. For example, Wernicke's area becomes active and uses lots of energy (which it gets from glucose) when we are reading.

Right now, your Wernicke's area is using more energy than any of the brain areas that surround it. If you are listening to music while you read, your auditory cortex is activated as well. By contrast, most of your motor cortex is "idling" and not using much glucose because you are sitting still.

MEASURING THE AMOUNT OF ENERGY THE BRAIN USES

Some diseases diminish the amount of energy that specific parts of the brain use. Some parts of the cerebral cortex in people with Alzheimer's disease use less energy than those same parts in normal people. Scientists have recently learned that parts of the cerebral cortex (although not exactly the same ones as in Alzheimer's disease) of cocaine addicts seem to use less energy than these same parts use in the brains of normal people. The decrease in energy use in Alzheimer's patients is the result of the death of many millions of cortical neurons. There is no evidence that cocaine kills neurons, and no one yet knows what causes the decrease in energy use in the brains of cocaine addicts, but the change seems long-lasting.

Ironically, two completely different interpretations could explain this energy decrease in the brains of cocaine addicts. First, information-processing in the affected parts of the cerebral cortex may be diminished. These brain regions just aren't working up to their potential because cocaine has changed them or has changed some other part of the brain that sends information to them. This is a reasonable interpretation because disease or injury typically reduces brain metabolism (energy use) in the affected area.

On the other hand, these parts of the cortex, which include portions of the frontal and temporal lobes, might actually be working more efficiently and, because of this, use less energy. This appears to happen with learning. When you are learning a new task, the brain regions that are needed for that task require large amounts of energy during the learning period. After the task has been learned, however, the same parts of the brain use much less energy to perform it. This second explanation doesn't seem to make sense for chronic cocaine addiction. In fact, most, if not all, scientists who study addiction think that the reduced energy use represents some form of damage or diminished ability of the brain. Whether it does or not and whether it is reversible, remains to be seen.

COUNTING THE BRAIN'S RECEPTORS

PET scans can also look at changes in the *numbers of receptors*. To do this, scientists put a radioactive tag on a special chemical that binds to the receptor under study. This approach has already provided evidence that cocaine addiction decreases the number of dopamine receptors in a part of the basal ganglia that is adjacent to and much like the brain reward system. The brain reward system is too small to be seen with PET scanning, but experiments with monkeys shows that it too is affected. However, although PET scans can identify brain regions involved in different functions and show how the activity of these regions changes under different conditions such as drug addiction, it is not always clear what those changes mean or what exactly causes them. Still, PET scanning and other similar techniques supply scientists with powerful means to study the brain while it is in action.

Therefore, when we talk about changes in the brain, we may mean:

- Changes that reduce the number of neurons in the brain

- Changes in the structure of neurons (such as their dendrites or axon terminals)

- Changes in molecules (such as receptors and second messengers)

- Changes in how genes work

- Changes in behavior that reflect brain changes we don't yet understand

- Changes in energy use

Although scientists can measure some of the brain changes that take place during certain conditions, many more remain unknown and mysterious. Sometimes, we do not even understand the meaning of the changes that we can measure. But every new piece of information gives us the opportunity to increase our understanding and ask more insightful questions as we search for ultimate explanations.

DRUG-INDUCED CHANGES IN THE BRAIN

Addictive drugs produce short-term changes in the brain, such as intoxication, and, if drug use continues, long-term changes such as tolerance, sensitization, physical dependence, withdrawal, psychological dependence, craving, and, finally, addiction.

In the rest of this chapter, we will explain the neural basis for intoxication, tolerance, sensitization, and physical dependence and withdrawal, and consider what these changes in the brain teach us about why people abuse drugs and progress toward addiction. We will explain psychological dependence, craving, and addiction in Chapter 9. It will be best to start by discussing the first changes in brain function that drugs cause, then to follow as drug use continues over time and additional changes occur.

SHORT-TERM CHANGES

Intoxication

Most people use drugs to get "high" or to relieve anxiety — sometimes for both reasons. Alcohol, marijuana, cocaine, heroin, and even caffeine generally produce at least mild euphoria, whereas alcohol, marijuana, heroin, sleeping pills, and even nicotine reduce or relieve anxiety. This combination of euphoria and decreased anxiety and, in some cases, alterations in perception, such as the changes produced by marijuana or LSD (lysergic acid diethylamide), constitutes the high that most drug abusers seek.

Let's look at the drug abusers we already met (see Chapter 2):

- Henry, the heroin addict, loved the rush, the intense feeling of euphoria that heroin produced.

- Sybil smoked cigarettes because they were at once mildly stimulating and relaxing.

- Neil and Michelle smoked marijuana because it produced mild euphoria, and because they thought the perceptual changes the drug produced were fun.

- Allison may have initially gotten drunk because she liked the mild euphoria she felt as she was drinking, but she also learned that drinking could reduce or even eliminate anxiety about sex.

- Chris used cocaine because he thought he needed the prolonged arousal it would produce, but it also produced euphoria, powerful feelings of well-being and increased confidence. He felt so good about himself while using cocaine that he thought his D paper would win him a Rhodes scholarship.

- Barry not only loved the initial rush he got from most of the drugs he used, he also liked the relief they provided from a persistent anxiety he felt most of the time.

Each of these drug abusers was responding to the acute effects of drugs, that is, the effects that appear first and are seen while the user is under the influence of the drug itself. We can describe people experiencing acute effects of abused drugs as being *intoxicated*.

Many people associate the idea of intoxication with the use of alcohol; it's a more formal way of saying that someone is drunk. However, it is reasonable to use the term "intoxication" to describe the acute effects of any drug when those effects are potent enough to impair emotional responses, perception, judgment, and performance.

Neil and Michelle ended up in a ditch because Neil was intoxicated on marijuana. When their friends tell the story to each other days later, they will say that Neil and Michelle must have been "stoned" or "really high" to do something so stupid.

People would not generally describe Sybil as being intoxicated on nicotine, because the dose delivered by each cigarette is too small to produce the profound impairments in behavior we associate with intoxication.

Allison violated a personal goal that was important to her because she was intoxicated on alcohol.

Chris's intoxication on cocaine led him to believe he was writing a brilliant paper.

Henry was willing to risk disappointing his wife yet again to experience the intoxication heroin gave him.

One acute effect of drugs is to alter judgment during intoxication. However, unless these alterations lead to a serious accident, produce a fatal overdose, or directly damage body tissue, the effects of the brain changes that intoxication produces are temporary and reversible. Thus, *intoxication represents a short-term alteration in brain function, not a long-term change in brain structure*.

The short-term, temporary effects of drugs on feelings, perceptions, cognition, and behavior occur because drugs cause changes in synaptic transmission. A good way to think about these effects is to consider the drug as if it were a neurotransmitter. Like neurotransmitters, drugs can cause effects *only* while they are present in the synapse. Also, like neurotransmitters, the length of time a drug acts depends on how quickly the brain can remove it from the synapse. As long as people use drugs only intermittently and infrequently, the drugs do not directly produce lingering or cumulative effects. Only when people start to use drugs more frequently does this situation change.

LONG-TERM CHANGES

When a person repeats drug use fairly frequently, changes in brain function, in addition to those that cause intoxication, begin to appear. What's more, these changes no longer disappear as soon as the body eliminates the drug. This shift in the brain's response is caused by the way its individual neurons adapt to the frequent presence of drugs, inducing persisting changes in the way neurons communicate.

Tolerance

Neurons change the way in which they receive and analyze messages as they adapt to drugs that are more and more frequently present in the brain. This adaptation is called *tolerance*, and it is the first enduring change that drugs produce in the brain.

A mother makes analogous physical changes in the way she holds her newborn as the baby grows and gets heavier. At first, she can easily cradle her infant in the crook of her arm. Later, she rests the baby on her shoulder. Still later, she typically rests the toddler on her out-thrust hip. Mom is still doing the same job of holding baby, but she adapts over time as the baby's weight increases.

When a person is tolerant to a drug, he or she requires more of that drug (a bigger dose) to get the effect that a smaller dose first produced. Most people find that their initial exposure to alcohol — one or two drinks — gets them high pretty quickly. If they drink only occasionally, perhaps once or twice a month, this will continue to be the case.

However, when a person drinks more frequently, as Allison did, the brain and body begin to cope with the increased presence of alcohol. Allison noticed this coping action as she gradually realized that she needed two or three drinks to get the same feeling of euphoria or relaxation that one or two produced earlier. Continued frequent drinking may increase the number required to three or four drinks for the same feelings to take effect. This effect is so well known that we have common expressions to describe tolerance to alcohol. Allison's friends began to describe her as someone who could really "hold her liquor," who had a "hollow leg," or who "could drink me under the table."

Tolerance to drugs such as alcohol, opiates, tranquilizers, and sleeping pills appears to have two components. First, the body becomes more efficient at metabolizing (disposing of) these drugs. This is called *metabolic tolerance*. Second, the cells of the body and brain change themselves to become more resistant to the effect of the drug. This is called *pharmacological tolerance*.

Metabolic Tolerance

Metabolic tolerance reflects the body's increased ability to eliminate a drug. The way the body eliminates alcohol is a good illustration of how this works. The body has two enzyme systems it can use to help get rid of alcohol: alcohol dehydrogenase and the microsomal ethanol oxidizing system.

The most important enzyme, which carries the major load when blood alcohol levels are not too high, is *alcohol dehydrogenase*. It breaks down alcohol into an inactive molecule, or metabolite, called *acetaldehyde*. Acetaldehyde cannot produce alcohol's effects on the receptors in the brain. The body contains a specific amount of alcohol dehydrogenase, and this amount correlates pretty closely to a person's body weight.

Alcohol dehydrogenase also works at a known, stable rate. The combination of these two factors is why we can predict how many drinks it will take a person of a given body weight to get drunk or how quickly the alcohol from one or two or three drinks will disappear from the bloodstream. Using this knowledge, police can determine whether a driver may have been driving drunk even several hours after an accident when blood alcohol levels are low. Because they know the rate at which alcohol dehydrogenase removes alcohol from the blood, they can calculate from how much of the drug is left to determine how much there was 30, 60, or 120 minutes earlier.

The liver contains most of the body's alcohol dehydrogenase, but the stomach also contains some of this enzyme. As a result, some of the alcohol a person drinks never gets into the blood, because the stomach metabolizes it. Men have more alcohol dehydrogenase in their stomachs than women. This is partly why men are less affected by a given dose of alcohol than women are, and it is why low-risk drinking levels are set lower for women than for men.

The other enzyme system that can metabolize alcohol is called the *microsomal ethanol oxidizing system* (MEOS). It is part of a large family of liver enzymes involved in metabolizing many drugs. Alcohol dehydrogenase takes care of most alcohol when there is only a relatively small amount in the blood. But as the blood concentration increases, the MEOS becomes active. In response to frequent large amounts of alcohol in the blood, the body can make more of the enzymes in the MEOS and metabolize alcohol more rapidly. However, the body's ability to increase the numbers of these enzymes is limited. After the body has produced a certain level of MEOS enzymes, it can't produce any more. So the amount of tolerance people can develop to alcohol, while impressive, is limited.

How Neurons Become Tolerant

There are enzymes such as alcohol dehydrogenase to metabolize other drugs such as cocaine or heroin. So, part of the reason people become tolerant to drugs is because their bodies make more of these specific enzymes and eliminate the drug more rapidly. But metabolic tolerance is only part of the tolerance people can develop. Not only can the body increase its ability to get rid of drugs, the neurons in the brain can reduce the effects of the drugs that do get there.

Although we do not yet know how neurons become tolerant to all drugs, scientists have identified two kinds of mechanisms that create this tolerance.

- Neurons can change the number of receptors they make.

- Neurons can reduce the receptor's ability to activate the second-messenger system (see Chapter 4) through which it works.

Let's examine how these changes might work. To do this, it will be useful to review a little physiology.

Homeostasis

In the last century, the physiologist Claude Bernard coined the term *homeostasis* to describe how the body manages to keep its internal environment relatively stable while the outside world changes. When we get cold, for example, our bodies undergo a number of alterations designed to conserve and generate heat. Shivering is one of those changes, and it increases the rate at which muscles use energy and thereby generate heat. This works the same way a heating system works to keep a house at a warm, stable temperature in the winter. You set the thermostat for the temperature you like. When the temperature in the house falls below the set point, the thermostat turns the furnace on, warming the house. When the temperature rises higher than the set point, the thermostat shuts the furnace off.

Our bodies maintain a stable temperature in a very analogous manner. A small group of neurons in the hypothalamus monitors our body temperature. If we get too cold, these cells send out commands that cause us to shiver, thereby warming us back up. Conversely, if we are too hot, other changes take place. One thing we do is perspire. As perspiration evaporates from our skin, it cools us. Both shivering and perspiring help us keep a constant body temperature in the face of a variable outside climate.

Our bodies regulate many aspects of our physiology, including our heart rate and blood pressure, the amount of oxygen and glucose in our blood, and even our balance, by following the same principle. As the world tries to push our physiology in one direction or another, homeostatic processes in our bodies work to keep it under control. Maintaining this control is what homeostasis is all about.

Adding or Removing Receptors

The principle of homeostasis helps to explain how neurons react to some of the changes in their environment. For example, a neuron that receives increased numbers of messages from another neuron that uses dopamine as a transmitter might begin to remove dopamine receptors from its membrane and make fewer dopamine receptors overall. Conversely, if the neuron received considerably fewer messages than usual, it might try to install more dopamine receptors in the part of its membrane facing the synapse. These changes help to keep the activity of the receiving neuron stable, maintaining the neuron's homeostasis. So, it is normal for neurons to modulate their sensitivity to their changing environment. Tolerance is an exaggeration of this normal neuronal response.

The removal or the installation of receptors or enzymes takes time and energy, and neurons do it only when a drug (or neurotransmitter) is at abnormal levels for a relatively long period of time. Once the neuron has reacted, the alteration in sensitivity to the message can outlast the presence of the drug. Altering the rate at which the neuron makes and degrades very complex molecules such as receptors happens over hours and days, far longer than the milliseconds required to send a message across a synapse. Moreover, these changes require the neuron to alter the way that some of its genes are working.

Tolerance to Cocaine

The principle of homeostasis gives us a framework that can help us understand tolerance to drugs. Cocaine, for example, increases the amount of dopamine in synapses. The principle of homeostasis would predict that the postsynaptic neuron (a presynaptic neuron sends the message; a postsynaptic neuron receives it) might try to reduce the dopamine message somehow. In fact, this appears to happen. There is now evidence that both people and monkeys who have self-administered cocaine for long periods of time have fewer dopamine receptors than do normal people (or monkeys, respectively). This decrease presumably helps the neurons that receive the dopamine message to regulate their activity back toward normal despite the abnormally large amount of dopamine in the synapse.

If you think about this in terms of the "boats and bikes" analogy (see Chapter 4), the dopamine boats would find that there just weren't enough dopamine docks to hold them all, thus limiting the number that could land and send their message. If this were to happen in the nucleus accumbens, cocaine might lose some of its ability to produce euphoria. And, in fact, human cocaine users report that the feelings of pleasure they get from cocaine diminish after they have used it for some time. Perhaps they have become tolerant to the euphoria-producing effect of cocaine by reducing the number of dopamine receptors in the nucleus accumbens.

Tolerance to Opiates

Tolerance to opiates develops through a different mechanism from that for cocaine. In the early 1970s when scientists first discovered opiate receptors in the brain, almost everyone suspected that tolerance to opiates would be the result of decreased numbers of opiate receptors. To everyone's surprise, this turned out not to be the case. In experiment after experiment, scientists tried to show that animals tolerant to opiates had fewer opiate receptors in their brains. But no one could show that the number of opiate receptors had changed.

Years later, scientists identified another mechanism. By then, they had shown that opiate receptors exert their effects within the cell by working through a second-messenger system (see Chapter 4). New experiments began to suggest that opiates somehow disrupted the links between opiate receptors and this second-messenger system. Thus, even though the number of opiate receptors didn't change in a tolerant animal, more receptors had to be activated to get the effect that opiates produced before tolerance had developed.

In terms of our boats and bikes model, tolerance to opiates results because the docks have fewer agents to collect the money from the boat captains. It doesn't matter how many boats dock, there simply aren't enough money collectors (second messengers) around to assist them all.

Here's another way to understand how opiates disrupt neurons' second-messenger systems. Imagine that you own a restaurant that gets crowded at lunch time. If your waiters and waitresses are efficient, you can earn lots of money. The waiters and waitresses are your second-messenger system; they relay orders from the customers to the cooks. You can make the most money by serving everyone quickly and getting them out of the restaurant so the tables can be turned over to the groups standing impatiently by the door. But, one day two of your waiters don't show up. The remaining waiters and waitresses can't keep up with the workload. People get fed more slowly, and you make less money. There simply aren't enough second messengers (waiters) to get all of the diners' orders to the kitchen so everyone can eat quickly. The noise from the clamoring crowd outside just isn't reaching the kitchen.

Variations in Degrees of Tolerance

Not all drugs produce the same degree of tolerance. That is, we can escalate the doses of some drugs much more than others. For example, some cancer pain patients, whose pain intensifies as their disease progresses, must continue to increase their dosage of morphine to deal with the pain. In doing this, they can become tolerant to doses of morphine that could kill a small group of normal people. Experience with these kinds of patients suggests that they have an almost limitless tolerance to the lethal effects of opiates.

By contrast, much less tolerance can be developed to the lethal effects of alcohol. Some tolerance clearly develops, but it is not possible to continuously increase the amount a person drinks without killing that person. This is undoubtedly due to the different mechanisms through which these two drugs act.

Variations in Rates of Tolerance

In addition, we become tolerant to different effects of drugs at different rates. Pain patients become tolerant to the *sedating* effects of morphine much more quickly than to its *analgesic* properties. This is fortunate, because their cognitive function returns to normal while the pain is suppressed. People also appear to become tolerant particularly quickly to the ability of drugs to produce euphoria. Experienced users commonly report that the pleasure they derive from drugs was greatest when they first started using, but then diminished.

Tolerance: A Problem for Drugs Abusers, Not Pain Patients

By itself, tolerance is not particularly significant. It simply represents one of the ways that the brain and body adapt to drugs. For example, when doctors give opiates to relieve pain, tolerance to the analgesic effects should not be a problem. It can be overcome by increasing the dose (although not all doctors are willing to do this). But tolerance is a real problem when people abuse drugs. This is because tolerance can set the stage for other more serious physiological changes. As drug users become tolerant to a drug, they increase the dose to keep getting the desired effect. If higher doses are used repeatedly and over long enough periods of time, physical dependence may develop.

Sensitization

Some drugs, including stimulants such as cocaine, can elicit a response that is just the opposite of tolerance. This response is called *sensitization*. That is, after using a drug repeatedly for a relatively short while, subsequent use causes an *increased* response, not the decreased response that results from tolerance. This is especially true for the anxiety-provoking and motor-activating effects of stimulants. Some people discover that they just get more and more anxious and restless every time they use cocaine. This happens even if they do not increase their dose. These people have become sensitized to these effects, and continued drug use can lead to periods of severe anxiety. Some even experience this with caffeine.

Sensitization is different from tolerance in another way. Instead of disappearing rapidly after drug use has ceased, sensitization can remain for weeks, months, or even years after a person has stopped using drugs. Because sensitization typically lasts for very long periods of time, it is clear that it is not simply the opposite of tolerance, but is the result of completely different mechanisms. Unfortunately, these mechanisms have not yet been clarified. At this time, it appears that sensitization may be the result of changes within neuronal circuits, not simply within single neurons, as with tolerance. Neurons that contain the inhibitory neurotransmitter gamma-aminobutyric acid (*GABA*) may be involved.

What makes sensitization particularly intriguing is that some scientists have suggested that a form of sensitization may lie at the very heart of some addictions. Think about the way that Alcoholics Anonymous directs former drinkers to behave. The injunction is to avoid alcohol at all costs. The justification for that strict guideline is that in many recovering alcoholics, a single drink will lead to a binge of drinking that the drinker is unable to control.

The alcoholic's response to alcohol is much larger than one might expect from someone who has not had anything to drink for a long time. It appears as if the response has been sensitized. This may be true for other drugs as well, because a single episode of drug use often leads to complete relapse into addiction for many abstinent addicts. While this concept is intriguing, the evidence that supports it is still scanty. We will have to wait to see whether or not it is true.

Physical Dependence

Tolerance and sensitization represent changes in the *magnitude* of the response to a given dose of a drug. The quality of the response, its characteristics, do not change, but the intensity of the response does. In addition, tolerance is easily reversible. Once drug use ceases, tolerance wanes over days or weeks. But, because tolerance forces drug users to increase the amount of drug they use, it sets the stage for *physical dependence*, another adaptation neurons must make to the frequent, excessive presence of drugs.

Physical Dependence Signaled by Withdrawal Symptoms

Being physically dependent simply means that when people stop using drugs, they experience uncomfortable withdrawal symptoms, which, in abusers, always include a strong desire to use the drug again. A person can become physically dependent on most, if not all, drugs of abuse. Day-to-day brain function does not appear to be altered, but from the point of view of an affected neuron, the change is profound.

The Cellular Mechanisms of Physical Dependence

Physical dependence reflects additional changes that the brain and body undergo as they adapt to drugs. Once physical dependence on a drug has developed, the brain and body literally need the drug to work properly. To understand what this means, think of the popular tug-of-war game many people play with their dog:

- You have a squeaky toy.
- Your dog wants it.
- The dog grabs it, pulls it.
- You pull back on it.
- The dog has to pull and lean back to exert more force.

- Your dog leans back so far that he needs you to pull against him to maintain his balance.

- If you let go of the toy, your dog will tumble over backward.

And so, once the body becomes physically dependent on a drug, neurons and other cells are in a relationship with the drug like the relationship described between the pet owner and his dog. This is another example of the principle of homeostasis. Acting through one or more specific receptors, the drug forces the cell to change its activity, becoming, for example, more inhibited.

Physical dependence on opiates illustrates this point. Opiates inhibit neurons, so when opiates are present, the cells fire off fewer action potentials. To keep their outputs within relatively normal limits, neurons that are constantly exposed to the inhibitory effects of opiates change other aspects of their physiology to make themselves more excitable. In essence, the changed physiology of these neurons is pulling against the way the drug is forcing them to behave. Once the drug is removed, these cells lose their "balance," just as the dog did. They then overshoot their normal level of activity and go too far in the other direction. Neurons that have been constantly inhibited by opiates become overexcited. The increased activity of these out-of-balance neurons results in the constellation of symptoms that make up the *withdrawal syndrome* (see a detailed discussion later in text).

Scientists have studied extensively the withdrawal syndrome that results from physical dependence on opiates. As with withdrawal from other drugs, the withdrawal syndrome produced by opiates is very much like the mirror image of a drug's primary effects. Opiates, among other actions, cause constipation, reduce coughing, depress respiration, and reduce anxiety. It is not surprising that the array of physical and psychological signs that make up the opiate withdrawal syndrome include diarrhea, coughing, hyperventilation (breathing too quickly and deeply), and anxiety. In addition, addicts experience intense craving for opiates (see Chapter 9 for a more detailed discussion of craving).

Adaptation to a Changed Environment

Like tolerance, physical dependence is not in itself a problem for the body. It is an *adaptation* to a changed environment (some scientists are now calling it a *neuroadaptation* because the primary site of action is the brain), but not necessarily a harmful one. However, in drug abusers, physical dependence becomes a huge problem for two important reasons. First, some drugs, such as alcohol and nicotine, are toxic or come mixed with other toxic chemicals. Long-term use therefore puts the user at a significant health risk. Second, physical dependence resulting from drug abuse can strongly reinforce further drug use. People who self-administer drugs to produce pleasure or reduce distress and anxiety find that avoiding the unpleasant (at best) symptoms of withdrawal provides a powerful motivation to keep using drugs. The negative reinforcement produced by alleviating withdrawal symptoms is at least as important in maintaining drug use as the positive reinforcement the drugs produce. Indeed, given that addicts report that the rewarding effects of the drugs they use have diminished, this negative reinforcement may be the primary mechanism that propels the drug user toward addiction and then maintains it.

The Proof of Physical Dependence

How can you tell when someone is physically dependent on a drug? Knowing drug use patterns can provide a good indication. If a person is drinking a bottle of whiskey every day, smoking three packs of cigarettes a day, or using heroin four times a day, he or she is almost certainly physically dependent. But there is only one way to know for sure. Drug use must be terminated. If signs of withdrawal appear after cessation of drug use, physical dependence has developed. Until withdrawal begins to occur, there are no physical diagnostic signs and no medical tests to indicate that a person has become physically dependent on a drug.

Physical Dependence — All Bad?

By itself, physical dependence does not affect the way a person functions. People who have been dependent on opiates for years to manage chronic pain function normally (that is, as if they were not taking the drug), even in demanding jobs. Similarly, former heroin addicts who are maintained on methadone can lead lives that are for all intents and purposes normal, except that they must get a daily dose of methadone to prevent withdrawal. So, taking on a physical dependence, or trading in an addiction for one, may not be a bad deal at all.

If managed properly in an addiction treatment program, physical dependence is consistent with very important and significant medical and social benefits. This is the assumption behind both methadone maintenance and nicotine gum or the patch. Total abstinence may be the best goal of drug abuse treatment, but it need not necessarily be the first or only goal.

Withdrawal

Withdrawal occurs because the cells of the brain and body not only have become accustomed to the presence of a drug, but have come to rely on it to maintain physiological "balance" in their activity. An addict goes through withdrawal when drug use stops, because the addict's brain (and body) has not yet adjusted to the *absence* of a drug that had been present for a long time. This adaptation to the absence of the drug takes place largely during withdrawal and can be measured by the diminishing intensity of the different symptoms of withdrawal over time. Withdrawal ends when cells have reached a new level of adaptation that allows them to function normally without the drug.

Whereas it may take weeks or months (depending on the drug, the dose, the pattern, and the intensity of use) for physical dependence to develop, withdrawal appears as soon as the effects of the last drug dose wear off. In that sense, withdrawal is almost immediate, which explains why heroin addicts need to shoot up every 4 to 6 hours and why alcoholics and heavy cigarette smokers (nicotine addicts) have a drink or smoke shortly after waking up. They have all begun to go into withdrawal and need to take a dose of their drug to alleviate or prevent the withdrawal symptoms.

Most of the signs of withdrawal are specific for the class of drug and depend on which parts of the brain and body that the drug affects. Withdrawal from opiates therefore differs from withdrawal from cocaine, and cocaine withdrawal differs from nicotine or alcohol withdrawal. As indicated earlier, if you know what effects a drug produces, you can pretty much predict what withdrawal will look like. The drug forces neurons to behave one way, but when it is gone, the neurons, like pendulums, swing quickly back the other way. Most withdrawal symptoms will therefore be the converse of the drug's effects.

Withdrawal From Depressants

Take as examples the depressant drugs, including alcohol. All these drugs first relax the user and lead to sleep at higher doses. Depressants induce sedation and sleep because they enhance the effects of the inhibitory neurotransmitter GABA by binding to its receptors.

GABA is found all over the nervous system. One of its roles is to ensure that no part of the brain becomes overexcited so that clear, sharp messages are sent. In this sense, the brain works like a cymbals player in a marching band. After clashing the two cymbals together, the musician quickly thrusts them against his body to stop the sound. This results in a clear, sharp note, not one that lingers and interferes with the music that is to follow.

Throughout the brain, excitatory neurons, which send messages to other brain regions, have inhibitory neurons as close neighbors and attendants. The excitatory neuron, called a *projection neuron*, sends its axon away from the local vicinity (the dopamine-containing neurons in the reward system, for example, project their axons away from the ventral tegmental area (VTA) toward the nucleus accumbens, cerebral cortex, and other parts of the brain). The neighboring inhibitory neuron, an *interneuron*, only sends its messages locally (interneurons in the VTA keep their axons within the VTA). The projection neuron and its inhibitory interneuron work as a team to control neural activity and to make sure that messages sent to other brain regions are sharp and clear.

Whenever the excitatory neuron sends a message out, it also sends a parallel, recurrent (returning) signal to its attending interneuron, exciting the interneuron. This interneuron in turn sends its short axon right back to the cell that excited it in the first place. There, it releases GABA to help turn off the excitatory neuron so its message doesn't linger to interfere with the subsequent message. Depressants enhance this effect, and when a person drinks or takes sleeping pills, the projection neurons must work harder to send their messages properly. In fact, the proper dose of a sleeping pill can stop many messages, which is what helps to induce sleep. Once neurons have altered their excitability to overcome the enhanced inhibition of long-term depressant use, the projection neurons become dependent on that extra inhibition to keep their activity in balance. This happens in people who are addicted to any depressant, including alcohol.

With that background, imagine what withdrawal from depressants must look like. Projection neurons that had worked hard to increase their excitability in the face of ever more effective inhibition, suddenly have that brake removed. In common terms, they lose it. They become extremely excitable and overactive. This results in increased anxiety, insomnia, and even seizures with overt convulsions. In fact, withdrawal from depressant drugs can overexcite the brain to such an extent that it can be life-threatening in people with severe addictions to depressants. Therefore, withdrawal from depressants must be carefully managed by a physician.

Other Less Dramatic Withdrawal Syndromes

Withdrawal from depressants and alcohol induce the obvious withdrawal syndromes with clear physical symptoms, even if they are not all as dramatic and dangerous as withdrawal from severe alcoholism or a long-term sleeping pill addiction. Opiates, too, produce a withdrawal syndrome, with many clear physical symptoms. But as miserable as withdrawal from opiates can be, people do not die from it. Nor does everyone who withdraws from depressants have seizures.

The intensity of withdrawal depends on the type of drug the person took, how long it was taken, and how much was consumed. People have argued for years about whether marijuana causes physical dependence. Using an antagonist to THC to induce withdrawal rapidly, scientists recently described a relatively mild set of marijuana withdrawal symptoms, including restlessness and insomnia.

Withdrawal from stimulants, including cocaine, has been difficult to define because it has few physical signs. What has become clear, however, is that withdrawal from stimulants is characterized by powerful drug craving.

Addicts who have frequently taken large drug doses over the course of a longstanding addiction are likely to have relatively severe withdrawal syndromes. Those who have used smaller doses for shorter periods of time will have less intense symptoms. The nature of withdrawal can also reflect the drug itself. The withdrawal syndrome that heroin produces is relatively brief and intense. Methadone, on the other hand, produces a less intense withdrawal, but one that lasts much longer.

Although withdrawal symptoms can vary with each drug, there is one withdrawal symptom that all drug addicts experience universally — craving. During withdrawal, addicts develop an intense craving for their drug. This craving is so powerful that it commonly leads to relapse into drug use even in addicts highly committed to stopping. In this chapter we have described the development of tolerance, physical dependence, and withdrawal in people who continue drug use. In Chapter 9, we will explain psychological dependence, addiction, and craving, and explore their meaning.

CHAPTER 9

HOW DRUGS CHANGE THE BRAIN TO PRODUCE PSYCHOLOGICAL DEPENDENCE, CRAVING, AND ADDICTION

How Chris Spent His Summer

Chris was anxious, excited, and elated that he had scored. He stabbed the elevator button repeatedly, pacing back and forth. He could hardly wait to get a few snorts of the incredibly pure powder he was gripping in his pocket.

All in all, it had been a pretty crummy year, and Chris wasn't often elated about anything anymore. Halfway through the academic year, the dean made him leave school. That "D" paper turned out to be the first of many, as "partying" replaced his study and career goals. Chris's grades fell so low that the dean told him the only way to avoid flunking out of college was to take the rest of the year off.

Next year, I'll show him, Chris vowed.

Then last month, Carolyn, his long-time girlfriend, dumped him. Screw it, she'd become a pain anyway, always nagging about his lousy grades, broken promises, and reckless outlook on life. Who was she to tell him how to live his life?

At home, his parents put relentless pressure on Chris to quit lying around the house and get back to school or get a job. Well, now he had a job — an office job. That "oughta shut 'em up for awhile."

Chris was feeling particularly lucky today, because not only had he scored big time, he also had found a new dealer, and the guy had great stuff! Chris's former dealer kept hassling him for the money he owed. Who needs that jerk anyway, he thought, as he rushed into the elevator, brushing past exiting passengers.

131

Each second of the trip to the 26th floor felt like an hour, and Chris, barely able to contain himself, tapped his foot and drummed his knuckles on the elevator wall all the way up. Finally, at the 26th floor Chris burst out of the elevator, raced down the hall to his office, and slammed the door as he rushed inside, flinging his briefcase, newspaper, and coat on the floor. He grabbed the coke kit out of his desk drawer, poured the sparkling white powder onto the kit's alabaster slab, scraped together a few lines of cocaine, and sucked those babies up his nose. Yessss!!

As soon as the euphoria dissipated and was replaced by nervous, edgy energy, Chris tried to turn his skittering attention to his next task — getting money for tomorrow's buy. Actually, getting money had become ridiculously easy, once he figured out a foolproof scheme. Chris had landed a job for the summer in Cleveland, where his uncle had gotten him a position as temporary bookkeeper for a nonprofit agency, which was one of his favorite charities. In addition, Chris's uncle had planned a summer trip to Europe and offered Chris his house rent-free if he would look after it.

Until Chris had devised his money-making scheme at work, this turned out to be a handy arrangement — he pawned useless things from his uncle's house to get money for coke. What the hell, he'd never miss them. But Chris had to admit he'd overdone it a little and would be in big trouble when his uncle got back. Yeah, well, I'll solve that problem later, he told himself. Maybe I'll arrange a fake burglary and blame the "thief" for the missing junk. Bunch of crap, anyway.

Last month, Chris had hit upon a *new* scheme, so he didn't have to pawn stuff anymore. One of his job responsibilities fit his needs nicely. Because he was studying accounting at college, Chris was given the job of writing the checks to pay the bills for the agency. It hadn't taken long to figure out how to type the checks, take them to the CEO for signatures, then run some of them back through the typewriter to replace the payee's name with his own, and deposit the "adjusted" checks into his account. Amazing what technology could accomplish!

Chris set up accounts at several banks to avoid suspicion and deposited checks and withdrew cash electronically from alternating banks. Although too preoccupied to keep track, he figured he'd appropriated about $25,000 so far, and nobody had even noticed. By the time they do, I'll be outta here, Chris promised himself.

On this day, Chris spent the morning paying bills, writing checks, and snorting cocaine. By lunchtime he had used up all his coke. He decided to pay his new dealer a visit during lunch, but first he'd have to "adjust" a few more checks and stop at a bank on the way. No problem, he told himself happily, as he headed down the hall toward the CEO's office with a fresh batch of checks. Piece of cake.

PSYCHOLOGICAL DEPENDENCE

With Chris, we see the essence of addiction: he has lost control of his behavior. Like all addicts, Chris primarily spends his time obtaining his drug, taking it, and being high (intoxicated). This activity overrides everything else that once was important to him — family, friends, school, work, hobbies, you name it. He has lost control of his drug use. And he has lost control of his behavior.

Addicts like Chris do things they would prefer not to do and would not do if they were not addicted. They may lie and steal from those they love to get money for drugs. They may become violent and even hurt the very people they care about most. They may cause wrecks when they are in their cars or put themselves at high risk for devastating illnesses that are easily avoided by people who don't use drugs. Drug addicts know they are harming themselves. They know they are harming those around them. Still, they do not stop using drugs.

Chris's addiction has progressed through tolerance and physical dependence to now include psychological dependence. Instead of using his natural abilities to cope with life's problems, Chris, like all drug addicts, has learned over time to take drugs to help him cope with stress, anxiety, depression, and other emotional problems. He has also typically used drugs to celebrate, so Chris also has learned to replace normal feelings of pride, pleasure, and satisfaction with drugs. This is the heart of *psychological dependence*. Chris has turned to drugs to provide pleasure and to avoid dealing with the real problems of life (and the emotions that these problems generate). The more he allows drugs to enhance, diminish, or replace normal coping mechanisms, the more psychologically dependent he becomes on those drugs. Unchecked, this progression ends when using drugs becomes the drug user's paramount way of dealing with daily living, as it has with Chris.

PSYCHOLOGICAL DEPENDENCE + PHYSICAL DEPENDENCE = ADDICTION

It is the eventual addition of psychological dependence to physical dependence that heralds the onset of addiction. Physical dependence is one of the necessary components of drug addiction, and almost everyone who becomes physically dependent on a drug because of drug abuse is at great risk for becoming addicted. But being physically dependent on a drug is not the same as being addicted.

Let's reiterate this point because it is often misunderstood: Physical dependence is a necessary component of addiction, but by itself is not sufficient to define addiction. Physical dependence must be accompanied by psychological dependence before we can say that a drug user is truly addicted. A drug abuser who is psychologically dependent believes that he or she cannot survive without the drug. Simply put, when asked about the drug, the addict says, "I must have it." In addicts, drugs fill a central role in life. On the one hand, obtaining drugs, using drugs, and being high take a lot of time. On the other hand, the addict has learned to use drugs to fulfill almost all needs, especially emotional needs.

Who Is Vulnerable to Developing Psychological Dependence?

The ease with which psychological dependence occurs depends, in part, on how mature the user is when he or she begins a drug-using career. People with a relatively strong sense of self, who have learned and practiced effective coping mechanisms and who have more experience with the vagaries of life in general, are more resistant — though certainly not immune — to the progression of psychological dependence.

Young People

Younger people, especially teenagers, are the most vulnerable to addiction. Because they have not yet developed and practiced effective ways of dealing with life's problems, adopting drug use as a way of coping with difficult feelings is attractive to those who start using drugs in their early teens. Adolescence is such a vulnerable age because in addition to having immature coping skills, adolescents are going through difficult physical and emotional changes produced by their own bodies. Unfortunately, initiation into drug use at a young age results in people who never get to learn how to live normally.

The Cost of Psychological Dependence and Addiction

Most successful adults have spent their adolescence growing up and actually learning how to be successful adults. Because they were at home and in school, they were in a relatively protected environment, where they could learn to deal with troublesome situations and negative or overpowering feelings. They sometimes succeeded and sometimes failed, did well and did badly, experienced new feelings, and learned how to control them as their bodies changed from juvenile to adult forms. All through this growing and learning period, young people get to practice their responses to life protected from the severe repercussions that are part of adulthood. Skipping school may result in no more than a reprimand. But skipping work can result in loss of a job. The stakes of such behavior are higher in adulthood.

Intensive drug use during adolescence, which should be devoted to growth and maturation, prevents, interrupts, and derails this important training. As a result, when people who became addicted as adolescents do get into treatment, they cannot be rehabilitated. They don't have skills that can be reclaimed or repaired. Instead, they have to learn life skills from scratch. This puts them at a considerable disadvantage in relation to their peers who have no such developmental delays.

Adults Are Also Vulnerable

Even though the learning that constitutes psychological dependence takes place more easily in young people, it can happen in more mature people as well. No matter when it occurs, the key is that the user learns to use drugs to "solve" problems. The solution may not be real to the rest of the world, but the changes induced in the brain by the drug are real enough to the user. Not only do drugs make the emotional distress of life's problems go away, but they also produce euphoria. Drug users figure that's not a bad deal. Eventually, the user learns to apply this package of pleasure and relief from stress not only when there is stress, but when there isn't stress. Why not? It's better than reality. Gradually replacing the experience of life with drugs is how psychological dependence develops: people learn it.

Are Pain Patients Drug Addicts?

Let's discuss one reason why we have been so insistent about.the difference between physical dependence and psychological dependence. If addiction were merely physical dependence, then people who are physically dependent on a drug for a medical reason would be drug addicts. But research has overwhelmingly demonstrated that they are not. Although it is possible to become addicted after being introduced to opiates through the treatment of pain, this almost never happens. In one study of 10,000 burn patients, none developed addiction. How can this be?

The big distinction between pain patients and drug addicts is that pain patients have lives that are full of things besides drug-taking, things like their families, jobs, and hobbies. These people, like most fully healthy people, focus on all these aspects of their lives and see the drug as the means (in this case, by alleviating pain) to be able to do those things. The addict, by contrast, sacrifices all these things just to use drugs. The psychologically dependent person knows this implicitly, if not explicitly, and seeks drugs with great intensity.

For the addict, drug use has become a career in itself and the very reason for being. Once psychological dependence has developed, most addicts feel that drug use has become, along with eating, drinking, and sleeping, a survival function. This is simply not the case for physically dependent pain patients, who gladly give up their opiates once their pain has passed.

CRAVING

Once people have become psychologically dependent, and therefore addicted, withdrawal of the drug has an even more profound impact on maintaining that addiction. After each dose wears off, withdrawal symptoms appear. The physical signs of withdrawal can be dramatic, but they differ from drug to drug. However, one aspect of withdrawal is common to all drugs of abuse. As addicts enter withdrawal, they feel the need to take more of the drug. They begin to crave it. This inevitably leads them to increasingly intense activity to acquire and use more of the drug. Most experienced addicts learn not only to recognize craving in its earliest stages, but to predict when it will occur. They act to alleviate craving long before it becomes powerful and consuming.

What is craving — that apparently mysterious and seemingly irresistible force that drives addicts to use drugs? The nature of craving is no mystery at all. It is easy to understand; it is just like hunger or thirst. The longer it has been since you last ate, the hungrier you get. Hunger is craving for food. Drug craving is hunger for drugs.

Hunger for Food versus Hunger for Drugs?

Despite the fact that we understand hunger and have no trouble justifying certain behaviors based on it, society has somehow placed drug craving in another category. Think about your responses to the following two scenarios.

1. A poor-looking young man steals a handbag. As soon as he gets away, he takes the money from the wallet in the handbag and uses it to buy food because he hasn't eaten for 2 days.

2. Another poor-looking young man steals a purse, but uses the money to buy a bag of heroin. He immediately shoots up because he has been in withdrawal for 2 days and needs a fix very badly.

In both cases, the stealing is wrong. But, although many people might have some empathy for the strong motivation to eat (the man was starving after all) that the first thief had, few would have a similar empathy for the equally strong motivation (he had intense drug craving) the second thief had. With the first thief, we might consider hunger for food to be a mitigating factor in judging the crime. With the second thief, hunger for drugs would be an aggravating circumstance, and we would condemn that thief even more strongly. Obviously, this distinction makes sense.

Hunger is a natural drive that helps us survive. Drug craving, by contrast, is something the addict produced in him- or herself while engaging in what began as an illegal, but voluntary activity. From this perspective, it is reasonable to consider that craving for food and craving for drugs are not at all alike.

Let's consider the similarities between hunger for food and hunger for drugs. Hunger for food arises from a physical need of the body. It is based in the biological processes that sustain our very life. All of us can understand this and find it reasonable to respond to hunger by seeking and eating food. But addiction derives its all-consuming, powerful hold on people because *it uses those same biological processes.* Hunger for drugs is based on the biology of the changed brain and arises from a physical need of the brain to maintain its current homeostasis. The motivation to use drugs, which addicts perceive as drug craving, emanates from the very same neural structures that give rise to the motivation to seek food, which everyone perceives as hunger. In the addict, the hunger for drugs has become a drive that is equivalent to and sometimes stronger than the hunger for food.

Even though we may not have personally experienced craving for drugs, we can appreciate something of what it feels like to have this kind of craving because we have all been hungry and thirsty. To understand this with even more immediacy, try to hold your breath for 2 full minutes. The desire you have to breathe after the first minute or so is a craving for air that you cannot resist. Your craving for air is craving in a clear biological context, and all you have to do to understand drug craving is to imagine how impossible it is for you to resist your craving for air. It is just as impossible for addicts to resist their craving for drugs.

"Just in the Mind?"

People have typically, and incorrectly, considered drug craving to be a "psychological" manifestation of withdrawal. This means that drug craving is somehow separate from a physical manifestation such as diarrhea, tremors, or runny eyes. The physical manifestations of withdrawal can be seen and measured by an outside observer. People believe that craving is experienced only in the mind. The separation of psychological phenomena, which occur in the mind, from physical events, which occur in the body, has long been a mainstream concept in Western thought. As a result, even though craving was well described decades ago, it somehow failed to attain the status of a "real" withdrawal symptom. People thought that the craving a recently abstinent addict experiences is "only in his mind," that he might even "be imagining it," that he could get over it "if he just had a little will power."

THE NEURAL BASIS OF CRAVING

Times have changed. Now, almost all scientists who study the brain are convinced that psychological states and events that we perceive in our minds are really the conscious manifestations of our brains at work. We see, think, feel, and move because of the work that goes on in our brains. From this perspective, craving is just as "physical" as any other symptom of withdrawal. It reflects a *state of brain function* brought about by drug withdrawal.

PHYSICAL DEPENDENCE MANIFESTED PSYCHOLOGICALLY: THE STORY OF COCAINE

We can see the implications of the failure to understand the true nature of craving with the cocaine epidemic of the 1980s. One reason why this cocaine epidemic became so severe was that we did not understand then that the psychological symptoms of withdrawal are as much a result of biological changes as any of the other physical withdrawal symptoms, such as diarrhea. At the beginning of the 1970s, most people believed that to be addictive, a drug had to produce physical dependence, as measured by the bodily symptoms of withdrawal. The overt and robust physical withdrawal symptoms produced by drugs such as alcohol and heroin were the models people (including scientists and clinicians) had in mind when they described physical dependence. However, cocaine produces very few of these kinds of symptoms. But, like other addictive drugs, cocaine does produce intense craving.

Still, the lack of clear physical symptoms associated with cocaine withdrawal led many people to conclude that it doesn't produce physical dependence. Because the presence of physical dependence was (and still is) considered a necessary component of addiction, any drug that didn't produce physical dependence simply could not be considered addictive. Many people therefore believed that cocaine was not addictive. In contrast to "hard" drugs such as heroin, which produce addiction, cocaine was considered a "soft" drug like caffeine, which doesn't produce addiction. It was safe (safer than heroin, anyway). In 20–20 hindsight, it is abundantly clear that this analysis was wrong.

The logic of the analysis that led people to believe that cocaine couldn't produce addiction because it didn't produce physical dependence was sound, but the main assumption — that cocaine doesn't produce physical dependence — was not. Cocaine does produce physical dependence, but people concluded the opposite because they did not understand that feelings, thoughts, and behavior are the products of the physical (biological) work of the brain (information processing), just as blood pressure and movement of blood through the arteries and veins are the products of the physical (biological) work of the heart (contractions).

Craving: A Physical Withdrawal Symptom

The brain is the organ of our mind (our consciousness) and of our behavior, just as the heart is the organ of our circulatory system. Thus, because psychological withdrawal symptoms such as craving are manifestations of brain activity, they are just as physical as physical withdrawal symptoms such as diarrhea, which is a manifestation of intestinal function. Just because a drug does not produce a withdrawal syndrome that consists of dramatic physical symptoms in the body does not mean that it is not addictive.

Addiction is a brain disorder. Whatever else they may do, after long-term abuse, addictive drugs produce physical changes in the brain that lead to the withdrawal symptom of craving. As uncomfortable and dramatic as bodily symptoms of withdrawal might be, when we ask what maintains drug use in addicts, or what causes recovering addicts to relapse into drug use, a large part of that answer is craving. Craving is a physical withdrawal symptom of the addicted brain.

WHY PSYCHOLOGICAL EXPLANATIONS FOR BIOLOGICAL PHENOMENA?

We have just argued strongly that psychological symptoms are really representations in our minds of physical, biological changes in the brain. But earlier we argued just as strongly that physical dependence isn't the same as addiction. To be addicted, one has to be psychologically dependent as well. How can we distinguish between physical dependence and psychological dependence on the one hand, and yet consider them the same (both physical) on the other? The solution lies in the insufficiency of our knowledge. Neuroscientists are convinced that brain activity underlies psychology, but they can't always describe the neural mechanisms involved in psychological events or states.

When the correct brain structures are active, the results are thoughts, feelings, and behaviors. But because we are just beginning to learn how the brain works, we cannot yet describe psychological phenomena in terms of brain mechanisms. We have yet to discover the neurobiology of almost any of the phenomena we use psychological terms to describe.

Even if we did have the required understanding, we would probably continue to use psychological constructs for describing our thoughts, feelings, and behavior. Psychological concepts do this so well because they transcend the neurobiology and focus attention on the most pertinent levels of analysis — the thoughts, the feelings, and the behavior.

Scientists, in trying to understand the world, typically reduce complex phenomena to their simplest components and mechanisms. But, although a complete understanding of the world might require this, it isn't always useful to be so reductionistic. For instance, you don't need to understand how your transmission works to know when and how to shift gears in your car. You don't have to understand the chemical composition of spices to cook a delicious meal. And you don't have to understand how your brain processes information about language to read.

Being able to describe psychology and behavior in terms of brain mechanisms will give us new insights into how we work. It will almost certainly be necessary if we are to develop new, more effective ways to treat diseases, including addiction. But if we are going to deal with addicts, we need terminology that is useful in describing their drug problem, knowing how to diagnose it, and knowing how to measure the outcome of treatment. No matter how much we understand about the brain mechanisms of addiction, we must always come back to the defining characteristic of addiction: drug-taking behavior. As a result, we use psychological terminology because it is appropriate. It has a tangible reality and demonstrable usefulness in everyday life.

PHYSICAL AND PSYCHOLOGICAL DEPENDENCE REVISITED

We have taken this digression to explain about the biological basis of psychological dependence and the need to use psychological concepts to define states that are really neurobiological. We have done this to clarify the reasons why our language sometimes confuses our understanding of abuse and addiction. Remember that both types of dependence, physical and psychological, are biological because drug-induced changes in brain function lead to both. Still, each type of dependence is different from the other because no matter how or why the drug got into the body, the presence of the drug by itself is sufficient to cause physical dependence, but not sufficient in and of itself to cause psychological dependence. Psychological dependence requires something else, and that something is how the drug is used and the motivation behind its use.

In chronic drug abuse, abusers use drugs repeatedly at high doses for long periods of time. The motivation is pleasure-seeking, avoidance of reality, and alleviation of stress (i.e., to avoid or alter the reality of life itself). That motivation can create not only physical, but psychological dependence.

It is remarkable that a major effect of abused drugs — the ability to produce addiction — depends on the drug's interaction with the user's frame of mind and behavior. Addiction, which we define in terms of behavior but explain in terms of changes in the brain, is due to an interaction of a drug, the user's motivation and behavior, and the user's brain. No wonder it is such a difficult problem to understand and deal with.

ADDICTION

Can We Define Addiction With Scientific Precision?

No. We have not assembled enough pieces of the puzzle to be able to have a precise definition of addiction, so for now we have to be satisfied with a working definition. The best working definition of addiction that we have is the loss of control of drug-taking behavior. When drug users reach a fully developed state of addiction, they feel compelled to use drugs, no matter what the consequences may be. A person's behavior changes as he or she becomes an addict, and it becomes radically different from what it was before drug abuse started. People who do not understand that the brain controls behavior would not ascribe the addict's behavioral transformation to changes in the brain, although this is in fact what has happened.

Although we do not know precisely which drug-induced changes in the brain cause addiction, we do know that long-term exposure to drugs alters brain function and structure. The striking changes in behavior that characterize addiction suggest that these differences are profound.

Even though we have argued that addiction is a brain disorder, defining it as the loss of control of drug-taking behavior focuses our attention on the key characteristic of addiction: the change in behavior. It tells us what to look for if we are trying to understand a person's drug use. What we can see and even measure is how often people take drugs, how much they take, and in what pattern. Just as we measure blood pressure to determine whether a person has hypertension, we can measure drug-taking to determine whether a person is an addict.

Different Patterns of Drug Use

Our working definition of addiction encompasses all classes of drugs, even though patterns of drug use differ for different kinds of drugs. Cocaine addicts, as Chris has become (see Chapter 2), typically go on binges with periods of abstinence in between. Therefore, their drug use pattern tends to be cyclical. Usually, alcohol addicts must drink every day, and nicotine addicts smoke dozens of cigarettes every day. The drug use pattern tends to be continuous. The working definition of addiction takes this into account, again focusing our attention on the key behavior.

Disadvantages of the Working Definition

Our working definition of addiction has some drawbacks. For one thing, it is not very precise. What does "loss of control" mean anyway? Once Chris has become addicted to cocaine, it's pretty easy to tell that he has lost control of his drug-taking behavior. People can see how far he will go to keep taking his drug.

On the other hand, loss of control of alcohol ingestion is not so easy to determine. If a person often chooses not to drink, but sometimes drinks too much, has he or she lost control? Or, how about someone who often drinks, but sometimes doesn't drink at all. Has this person lost control? Or, does loss of control mean almost never managing *not* to drink (continuous drinking)? This lack of precision is paralleled by a failure to indicate how abusers change into addicts. What is happening that is different? Our working definition, focusing as it does on behavior, is silent on this issue as well. For scientists who want to define precisely and measure accurately, this is unsettling.

Neuroscientists are working to create a definition that refers to a particular change or set of changes in the brain. They would like to be able to talk about a change in receptors, or second-messenger systems, which could be measured with a PET (positron emission tomography) scan or another diagnostic test. They would like to be able to say something like "a person with more than a 50% decrease in dopamine receptors is a cocaine addict." But we don't know enough to do that yet, and scientists therefore accept a definition with the kinds of drawbacks our working definition has while continuing to work to improve it.

Language Limitations

We can see that the weaknesses of our working definition of addiction reflect what we don't yet know or can't measure. These weaknesses also reflect something else: the limits of our language. Our language creates categories, carved out of our lives. For instance, boxing has rigid weight classes. If a boxer weighs more than 190 pounds, he is a heavyweight; if he weighs less, he is a light heavyweight. But most of the real world is a continuum of subtle differences (the difference between 189 and 190 pounds is trivial, except in boxing). Similarly, the borderline between being a long-term drug abuser and being an addict is not clear.

The distinction between nonaddicts and addicts can be compared with the changes that separate the various parts of the end of the day, the changes from afternoon to sunset, to dusk, to night. These changes are apparent after they have occurred, but not while they are happening. So, our definition of addiction, which depends on the key behavioral characteristic of addiction, reflects the uncertainty that exists in defining and measuring that characteristic with precision.

THE NEUROBIOLOGY OF ADDICTION

What we really want to know is just what is different about the brain of an addict. This is a very complicated question, and the answer is still mostly shrouded in mystery. It is not even clear that we have formulated the proper questions or that our technology can help us answer these questions even if we ask them properly. So let's first examine why understanding even the knowledge we do have is difficult.

Trying to Measure the Problem

Part of the problem in understanding the neurobiological basis of addiction is that the various components appear sequentially. It is possible to study tolerance directly because it appears first. But, because tolerance develops before physical dependence if you want to understand physical dependence, you must study people or animals who are both tolerant and physically dependent. Separating the brain changes that are responsible for tolerance from those that are responsible for physical dependence is no easy task. When it comes time to study addiction itself, we must take observations from human addicts and then try to distinguish the contributions made by tolerance, physical dependence, and psychological dependence.

Consider an analogy to better understand the problem. What makes a star athlete? Star athletes must have outstanding physical abilities. They must train with great dedication to develop those skills. But the most successful athletes also have a certain set of psychological qualities: determination, ability to concentrate, leadership, playing best when it counts most, and more. So, if we were to take one of these athletes at the peak of his or her career and ask what makes this person a star, how would we evaluate the contributions of these various factors? If we wanted to define a star, should we look for the skills, the training, the determination, or the ability to concentrate. If we see some of those traits in a person, can we conclude that he or she is a star? For now, the answer has to be: it depends. It depends on what other characteristics are also present. So, just as outstanding athletic skills are not all that it takes to make a star, neither is physical dependence sufficient to make an addict.

What this example demonstrates is that the simple presence of a particular trait does not necessarily mean that trait is the key or most important one. We may be able to say that cocaine addicts have fewer dopamine receptors than normal people, but we do not know whether this decrease in dopamine receptors is the key trait that defines cocaine addiction. We do not even know whether it is a cause or result of the addiction, or totally unrelated to it. As a result, there is no neurobiological definition of addiction. There are, however, a number of clues. Let's lay them out as best we can.

Clue #1. The Brain Reward System: The Neural Substrate for Motivation

Because cocaine addiction has recently been a tremendous social problem, let's use cocaine as a model system to study the neurobiology of addiction. Many neuroscientists suspect that cocaine addiction involves changes within the brain reward system because it is intimately tied up in the response to drugs and also because it is the neural substrate for motivation. As we have seen, chronic exposure to cocaine produces changes in the neurons that use and respond to dopamine. It is not clear whether these changes are related to tolerance or dependence, or both or neither, but they do occur.

Clue #2. Differences in the Brains of Cocaine Addicts

Dopamine Neurons Make More Reuptake Pumps

If one combines all the available human and animal data about cocaine (which is not an easy thing to do), it is possible to make some reasonable conclusions about what is different about the brains of cocaine addicts. In cocaine addicts, the neurons that release dopamine appear to make more reuptake pumps than normal. In addition, changes seem to occur in one of the key enzymes that makes dopamine.

Neurons Make Fewer Dopamine Receptors

Across the synapse from the axon terminals of these neurons, the neurons receiving their message make fewer dopamine receptors. In both humans and monkeys, this reduction in the number of dopamine receptors lasts for at least many months after drug use has stopped.

Certain Parts of the Brain Use Less Energy

In both humans and monkeys, cocaine causes certain parts of the brain, especially parts that are related to the brain reward system, to use less energy. In humans at least, this change seems to be rather long-lasting.

Clue #3. Differences in the Brains of Opiate Addicts

Changes in Second-Messenger Systems

Chronic exposure to opiates produces changes in the links between opiate receptors and their second-messenger systems, and these changes develop along with the development of tolerance. In addition, scientists have linked physical dependence on opiates to changes in two specific neural structures outside the brain reward system. In fact, the brain reward system appears to have nothing to do at all with physical dependence (as determined from physical symptoms) on opiates.

Changes in the Locus Coeruleus

One of the neural structures involved in physical dependence is called the *locus coeruleus* ("blue place"), because it has a slightly bluish color in a fresh brain (see Fig. 3.2). It is the source for the entire brain of a specific neuro-transmitter, norepinephrine. You may have heard of norepinephrine as part of the "fight-or-flight" response. This response occurs when we are faced with danger or severe stress. As part of the fight-or-flight response, the adrenal gland and the sympathetic nervous system release norepinephrine and its close chemical cousin, epinephrine. The norepinephrine in brain structures such as the cerebral cortex, thalamus, limbic system, and hypothalamus all comes from the locus coeruleus, which plays a role in controlling the sympathetic nervous system.

The locus coeruleus is involved in a variety of functions, including attention and arousal. Its role in regulating the sympathetic nervous system becomes apparent during withdrawal from opiates. Locus coeruleus neurons contain opiate receptors, and like neurons with opiate receptors elsewhere, opiates inhibit these neurons. Then, during opiate withdrawal, these neurons are released from inhibition and become very active. As a result, the sympathetic nervous system also becomes very active. Because of this, many of the symptoms of opiate withdrawal, such as dilated pupils, increased heart rate and blood pressure, and chills, look like part of the fight-or-flight response.

Changes in the Periaqueductal Gray Matter

The other neural structure involved in physical dependence on opiates is called the *periaqueductal gray matter*. When scientists inject morphine directly into this structure in rats, the rats become physically dependent on opiates. But rats will not self-administer opiates into the periaqueductal gray matter. So, even though it appears that the periaqueductal gray matter is involved in the development of physical dependence, it is not at all clear what role it actually plays.

Clue #4. Morphine and Structural Changes of Neurons

A remarkable finding has recently been published. A group of scientists working with rats report that chronic exposure to morphine induces changes in the physical structure of the dopamine-containing neurons that are part of the brain reward system. Neurons, like other cells, are essentially bags of chemicals. To hold themselves together and support their dendrites and axons, neurons make special proteins that erect frameworks for the cell body, dendrites, and axons. Reasonably enough, these are called *structural proteins*. However, when morphine is in the body, dopamine neurons make fewer structural proteins, their cell bodies appear to be smaller than normal, and their dendrites are not as long as normal. It is not clear what this means, but it adds a fascinating new dimension to the study of drugs.

Most research effort now is focused on molecules within neurons, such as receptors, second messengers, enzymes that synthesize transmitters, and reuptake pumps that actually participate in mediating the effect of the drug. An alteration in structural proteins indicates that drugs can have much broader effects on cells.

Missing Clues About Effects of Other Drugs on the Brain

Scientists have not yet developed information about the neurobiological effects of the chronic use of drugs such as marijuana, LSD, and phencyclidine (PCP). So, this relatively short list of brain changes that could potentially underlie addiction and help us define it more precisely tells us three things.

1. We are just beginning to acquire knowledge about the neurobiological basis of addiction. It is a relatively new field that only attracted significant societal and scientific attention with the onset of the cocaine epidemic of the 1980s.

2. The tools needed to reveal the brain changes have only recently become available, and even more sophisticated tools remain to be developed.

3. The scientific community can be highly responsive to the needs of society. Before 1985, most research on drugs focused on opiates. Since the middle of the 1980s, cocaine research has grown enormously. A smaller group of scientists study marijuana. Not only is there a widespread perception that marijuana is not nearly as dangerous as cocaine or heroin, but animals will not self-administer it. So its effects are not easily evaluated.

OUTCOME: LOSS OF FREE WILL

When the changes in the brain of the drug abuser become so profound that he (or she) can no longer control drug-taking behavior, he has become an addict. When confronted with drugs, he has lost his power of choice, the very essence of who he is: he has lost his free will. Instead of being free to choose what to do at any given time, he has developed a brain disorder that compels him to keep seeking and taking the drug to which he has become addicted.

Viewing addiction as a brain disorder does not mean that the addict has no ability to make choices. This is clearly not the case. But it does mean that the ability to make choices is altered. No one has yet defined the extent of this change in the addict's ability to make free choices, and we are not yet close to being able to ascribe such a complex process as decision-making to a particular part of the brain, so again we fall back with some uncertainty on our imprecise language. In this case, we have a relatively large selection of words that more or less convey what has happened. We could say that an addict's ability to make choices about drug use is impaired or that it has dwindled, become constrained, or enfeebled, weakened, diminished, or degraded. All these terms refer to an aspect of the changes in decision-making ability that has occurred. The ability to choose freely has been altered because long-term, repeated, high-dose drug abuse has changed the addict's brain.

There might be an alternative explanation for the failure to stop using harmful drugs. Let's consider this. Perhaps the ability to make sound decisions hasn't been altered after all; perhaps free will is not impaired. What if the ability to make decisions is left intact, but the person's judgment as to the importance of the reasons for such decisions has become distorted or reordered? How might this work?

Like this: heroin addicts describe their drug craving in much the same way that a sailor who has been stranded on a lifeboat with nothing to drink for several days would describe his craving for water. Of course, both the craving addict and the thirsty sailor could choose not to satisfy their craving when heroin or water eventually becomes available. We have little problem understanding the compulsion that the sailor would feel to drink, but we can't seem to find the same level of understanding of the addict's compulsion to "shoot up." Addicts use drugs even when there are clear medical, social, and legal reasons not to. Thus, their motives must be strong indeed.

Drug Use Becomes a Survival Need

From the perspective of motives, the addict's drug use does make "sense." That is, if it makes "sense" from a biological perspective to drink when you haven't had water for several days, it also makes biological "sense" to shoot up when your craving for drugs makes you feel the same way. The only difference is the substance, not the reality of the feelings the brain generates.

And this is the key: some of the same brain structures that create thirst — the craving we feel for water — create the craving for drugs that an addict feels when he or she goes into withdrawal. The parts of the brain that receive the message that something is being craved do not need to know, or even care about, what that substance is. Their job is to set priorities about what to do ("Is satisfying this craving more important than something else I might do?") and then give the command to accomplish the task at hand.

We can see this with monkeys that have been repeatedly exposed to cocaine. Given the choice between food or water on the one hand or cocaine on the other, the monkeys have little problem with this choice. They take the cocaine. We presume that they are responding purely to the strength of their motivation to get the drug in relation to their motivation to eat or drink. The monkey is behaving as a simple biological decision-maker, measuring drives for their survival value and giving the highest priority to the most intense craving at the moment.

However, humans are self-reflective in a way that we assume other animals are not. Thus, we not only make a biological comparison of our need for drug versus water (or drug versus no drug, or drug versus going to work), but we can add additional factors to the logic of that decision. These factors may include the knowledge that no matter how good using drugs makes us feel, drug use is bad for us. We are good at this kind of decision-making. So good, in fact, that our written laws and unwritten social conventions expect us to make decisions that can, under appropriate circumstances, go against our own immediate biological best interest. For example, when the Titanic was sinking after hitting an iceberg, the men put the women and children in the lifeboats at the cost of their own lives.

Drugs Diminish the Role of the Cerebral Cortex

Addicts don't act like rational humans because drugs diminish the role of the cerebral cortex — the very brain region that gives us our ability to evaluate a variety of contingencies and make reasoned choices by weighing them appropriately. In fact, all addictive drugs activate the brain reward system, which is situated to control behavior through primitive but powerful conditioning mechanisms that don't even need the cortex. At the same time, many drugs, especially depressant drugs, muffle the influence of the cortex by directly inhibiting its activity. As activity in the cortex is reduced, reason and higher learning are removed from the behavioral control process, and emotions and primitive motivations are released, as we saw most pointedly with Todd, Allison, Jennifer, and Megan, who wrecked their car while driving drunk.

Nevertheless, whether drug use impairs free will, or distorts the importance of the reasons for exercising it, the outcome is the same: the very ability to choose that a person exercises to initiate and continue drug use is gradually exhausted and ultimately destroyed by a brain disorder called addiction.

CONCLUSIONS

Right now, we have a reasonably good behavioral definition of addiction. Neuroscientists believe that we will one day be able to describe the brain mechanisms that lead to and underlie these addictive behaviors. At that point, neurobiology will have greatly enriched our understanding of addiction by revealing the brain mechanisms that underlie both observable behavior and psychological events perceivable only in the mind of the addict. We may even dispose of the concept of psychological dependence, even though it is now the most useful way of describing the difference between physical dependence and addiction. It is, after all, a theoretical construct that became necessary once it had become apparent that physical dependence is not addiction. If psychological dependence is indeed as biological as we have argued, we should one day be able to say that drug-induced changes in brain systems X, Y, and Z mean that a person is physically dependent, but that a person is not addicted (that is, also psychologically dependent) unless brain systems T and U are also changed.

For the time being, drug addiction is best viewed as the loss of control of drug-taking behavior. The loss of behavioral control is due to a brain disorder that results from long-term self-administration of drugs taken to get high. This brain disorder, like many others, is expressed through altered behavior, and we finally interpret it in a social context. The development of addiction may be influenced by the drug, the patterns and motivations for drug use, individual differences among drug users, and environmental contingencies that may change over time.

Treating addiction involves detoxifying the addict and then replacing all the maladaptive behaviors the addict gradually acquired while addicted with productive behaviors that will allow him or her to live life successfully. Although this is a daunting task, treatment for drug addiction is often successful. We will describe how treatment works in the next chapter.

CHAPTER 10

INTERVENTION AND TREATMENT

*P*erhaps the most difficult thing to understand about addiction is that many addicts do not want to stop taking drugs. Even though it becomes obvious how sick they are and that they are destroying their lives, addicts either cannot see this or don't want to see it clearly enough to do something about it now (rather than "later"). They have found a way to deal with their world that "works" for them. As they gradually learn to use drugs to cope with everyday living, the drugs they take become substitutes for the things addicts are unable to obtain in life, things most of the rest of us take for granted.

Instead of dealing appropriately with negative feelings so they can negotiate their way through life to get what they want, addicts have learned to avoid life by taking drugs. Drugs make them feel good or less bad or, by the time they are addicted, at least normal. They like these feelings. Even though addicts have lost control of their behavior and do things they would not normally do, most do not want to give up drugs.

As explained throughout this book, the persistent use of drugs creates biological changes in the addict's brain. These changes drive the persistent need to continue using drugs — no matter what. Consequently, addicts rarely enter treatment without pressure because, from their point of view, treatment means having to give up the one thing that is most important to them. Why would they want to do that?

INTERVENTION

Not understanding the basic premise of addiction — that structural changes in the addict's brain drive the addict to keep taking drugs — has led to misunderstandings, misconceptions, and myths about treating drug addiction. One of the most destructive of these misconceptions is that there is nothing you can do to help an addict until he or she wants help badly enough to ask for it. This misconception is based on the belief that the addict must "hit bottom" before anyone can do anything about the addiction. This is simply not true. This myth has grown out of a lack of understanding of the mechanisms and dynamics of addiction. It is time to put this myth to rest, because it has contributed to needless suffering among addicts and the people who love them.

We now know that addicts can be helped long before they hit bottom. The people who are closest to addicts — their families and friends — do not have to stand by helplessly watching the addict slowly self-destruct. They can *intervene* before the addict sinks anywhere near the "bottom" and get the addict into treatment.

The process of recognizing what is happening to a person who has become addicted to drugs and interrupting the addiction to get the addict into treatment is called *intervention*. Treatment helps the addict recover. Intervention gets the addict *to* treatment. Family interventions constitute only one form of intervention; there are also workplace interventions, legal interventions, medical interventions, and others.

Workplace Interventions

Although drug addiction can become all-consuming, some addicts do manage to work. Inevitably, however, their addiction leaves clues in their work performance and interactions with coworkers. Many companies have Employee Assistance Programs (EAPs), which are designed to spot the troubled behavior that can signal addiction — increased absenteeism, decreased productivity, increased accidents on the job, and so on. These programs help employees solve whatever is causing the changes in their job performance. If the problem turns out to be addiction, EAPs have a tremendous advantage in being able to intervene successfully. Even though the last thing the addict wants to do is change his or her behavior, these programs can get the addict's attention in a way few others can. This is done by making it clear that the addict has a choice: Enter treatment and we'll pay for it, or lose your job. Such a choice is frequently successful in forcing addicts to look at their drug-taking behavior, recognize that it is out of control, and enter treatment to regain control of their behavior.

Legal Interventions

Another kind of intervention occurs when addicts break the law and are convicted of committing a crime. Judges can often offer addicts the same kind of choice employers can: Enter treatment, or go to jail. Again, the shock of such a choice compels many addicts to deal with their out-of-control, drug-taking behavior and get help. We will learn more about legal interventions when we see how Henry finds his way to treatment.

Medical Interventions

Still another kind of intervention is conducted by doctors, rather than employers or courts. This kind of intervention usually occurs when an addict is seeking treatment for what he or she believes is an unrelated medical condition (i.e., a smoker who has a heart attack) and the physician intervenes by advising the patient that the best chance for full recovery from the more obvious medical condition is to stop drug use (i.e., the doctor tells the patient to quit smoking). Sybil's entry into treatment occurs through a medical intervention, as we shall see a little later in this chapter.

Doctors not only intervene in patients' addictions when they are facing life-threatening diseases resulting from their addictions, but they also intervene in other ways. Most doctors now screen for alcohol, tobacco, and drug use during annual checkups and talk with patients about what they report. If you smoke but haven't suffered any health consequences yet, your doctor will probably encourage you to quit and recommend that you try nicotine gum or a nicotine patch. If you drink too much alcohol or use illicit drugs, your doctor will most likely warn you of their dangers, refer you for an assessment for addiction, and recommend treatment.

Doctors also intervene when people are taken to emergency rooms with drug- or alcohol-related injuries or when expectant mothers go for checkups and reveal they are drinking alcohol or using other drugs. Pregnant women are also screened for drug abuse and referred to special clinics, where they can be treated for addiction and also obtain care for their other children while they undergo treatment. Unfortunately, too few of these facilities are available to treat all the women who need help.

Family Interventions

Another common form of intervention is conducted by the addict's family, as is the case with Barry and his family. Let's look at Barry (see Chapter 7). The last thing he wanted was treatment for his polydrug addiction. It hadn't even entered his mind. But his raid on the family's bank account forced Barry's parents to recognize that he needed help. They planned a family intervention.

Barry's Intervention

"Come in, Barry, and sit down," Barry's father Richard said when he saw his son coming downstairs. "We want to talk with you."

Barry wasn't sure what was up, but clearly something was afoot. He'd heard people gathering downstairs while he was getting ready to go out for the night and wondered what was going on. Seated in the living room were his mother, his sister Lisa, his two best friends, Neil and Michelle, and some guy his father introduced as Dr. Jenkins. He didn't like the looks of this, but it was too late to escape. As soon as he sat down, Dr. Jenkins began talking.

"Barry, all the people in this room are deeply concerned about you. Each one loves you, and each one has been watching you turn into someone they no longer recognize. They've been meeting together for the past month to figure out how to help you."

"I don't need any help," Barry blurted out, immediately on the defensive.

"Let's talk about that," Dr. Jenkins replied. "Lisa, why don't you begin."

Barry's family and friends spent the next several hours going through the painful process of a family intervention. At first, Barry was shocked by their assertions that anything was wrong. Then he became angry and abusive and tried to blame everyone else for his problems. But his family kept talking with him, refusing to take his bait, and slowly breaking down his denial. Ultimately, they forced him to see how his behavior had changed over the past year. They finally got him to admit he was addicted to drugs and needed help.

Family interventions are never easy — for the addict or the family. They almost always bring painful feelings to the surface, but they also bring the addict's self-destructive behavior to an end and move him or her into treatment and recovery.

Even First Families Intervene

Addiction can happen in any family, even "first families." Both famous and not so famous people have been treated at the Betty Ford Center. The center was founded by former First Lady Betty Ford, who became addicted to alcohol and prescription drugs while her husband was President. Shortly after the President left office, the Ford family intervened in Mrs. Ford's addiction and insisted that she enter treatment. They were willing to risk public humiliation out of their concern and love for her. What they got instead was public admiration and support for their courage. Nevertheless, they couldn't have predicted the public's reaction when they made the decision to act.

Mrs. Ford not only admitted she had a problem and entered treatment, but succeeded in recovering. She established the Betty Ford Center to help others who find themselves in similar trouble.

Parents' Universal Response — Guilt

Barry's family conducted an intervention similar to that of the Fords. Barry's parents, Cynthia and Richard, learned about intervention by calling a local drug prevention organization recommended by a friend in whom Cynthia confided. When they discovered that their child was addicted to drugs, it was overwhelming to Barry's parents, as it is to most parents in the same situation. They experienced an intense sense of guilt, as if it were somehow their fault. Generally, parents with drug-addicted children feel severely isolated, as if they are the only parents who ever experienced such a problem with a child. Like Cynthia and Richard, parents must overcome these feelings before they can act.

The picture changed for Cynthia and Richard as they reached out hesitantly for help. They discovered that they weren't alone. It turned out that several families within the broad network of people they knew professionally and socially were having similar experiences; they just hadn't publicized them.

Final Preparation: Selecting a Treatment Program

A successful intervention ensures that the addict enters a treatment program at its conclusion. One way to do this is to preselect a treatment program and enroll the patient in advance of the actual intervention. Cynthia and Richard visited a number of drug-free, inpatient, and outpatient treatment programs and selected one they felt best fit Barry's needs and their budget. They also obtained the services of Dr. Steven Jenkins, an intervention counselor, to help the family confront Barry and persuade him to enter treatment. Cynthia, Richard, Lisa, and Barry's two closest friends met weekly with Dr. Jenkins for about 1 month to prepare for the intervention.

Purpose of Preparatory Intervention Meetings

The purpose of the preparatory group meetings before the intervention takes place is threefold:

- To confront and break down the denial and enabling of family members and close friends

- To prepare the group to confront the addict, break down their own denial, and convince the addict that he or she is in trouble and needs help

- To prepare a plan that will conclude the intervention by placing the addict in a specific treatment program at the close of the intervention.

Denial

The intervention process is painful for everyone who participates in it. One reason is that, initially, it is rarely clear to the people who prepare for an intervention that drug addiction is what they are dealing with. They know something is wrong with the person they are concerned about, but their own ignorance about how addiction works and their inability to believe that it could be happening to someone they love prevent them from identifying the problem.

An intervention counselor guides family members and friends who will participate in the intervention to share what each one has observed about the addict. Almost inevitably, they discover that each person had been "keeping secrets" about the addict's behavior, mistrusting his or her perceptions, refusing to see that drugs could be at the heart of the behavior changes. This inability to see that a loved one is involved with drugs is called *denial*.

Most people do not want to see that someone close to them is even using drugs, let alone addicted to them, because seeing addiction means they have to deal with it. However, as family members and close friends share their observations, it becomes painfully clear that the person they are concerned about is addicted to drugs and will only get worse if they don't intervene.

Enabling

Intervention group members also become aware of the many ways they have unconsciously *enabled* the addict's drug-taking behavior to continue. In essence, they have made excuses for behaviors they would not normally tolerate and have ignored other behaviors. For example:

- A wife ignores her husband's increasingly unstable behavior, accepts his nightly drunkenness, allows his occasional physical abuse of her to go on — and continues to stock the liquor cabinet rather than risk losing the only financial support she has for her children and herself....

- An inner-city mother living in public housing never questions her son, who has no visible means of support, about where he got the money to buy her a new stove and refrigerator for her birthday, nor does she ask where the puncture marks, scabs, and bruises that regularly appear on his arms come from. Even though her son is a grown man, she continues to cook for him, launder his clothes, give him shelter, and buy him the things he needs in everyday life. . . .

- A suburban family repeatedly bails their son out of trouble, hiring lawyers and paying court costs, fines, and traffic tickets without ever asking themselves why these unlawful events happen to their son so often. . . .

After beginning the process of breaking through their denial and recognizing their enabling behaviors, family members and friends, with the help of the intervention counselor, are ready to confront the addict and persuade him or her to enter treatment.

TREATMENT

Just as there are many different kinds of interventions, there are also many different kinds of treatments. Although each treatment uses a different approach, each is designed to help the addict recover from addiction.

By the time a person has become addicted, significant changes have occurred in the brain. To recover, addicts must first get the drug out of their brain and body. This is called *detoxification* (removing the poison). It generally takes from a few days up to about a week.

The more difficult part of recovery takes much more than 1 week. Addicts must give their brain time to readjust to the absence of drugs and then learn behaviors that will allow them to stay off drugs. The addict must go through withdrawal, extinguish the unconscious cues that provoke drug craving, unlearn the destructive behaviors acquired when becoming addicted, replace them with constructive behaviors, and overcome the memories that reinforce drug use. If the addict became addicted before acquiring the life skills, social skills, and job skills necessary for successful living, he or she must also learn these skills to sustain recovery and become a contributing member of society. This combination of detoxification and therapy, and, in some cases, skill acquisition, is called *drug treatment*. It comes in various forms, including:

- Drug-free, inpatient, chemical-dependency treatment

- Alcoholics Anonymous and other 12-step programs

- Therapeutic communities

- Medical approaches that involve providing medications that substitute for or block the effects of the drug addicts are addicted to

Barry's Treatment

Drug-Free Chemical Dependency Program

Barry's family selected for him a drug-free, chemical-dependency treatment program that offers short-term, residential treatment. This kind of treatment involves detoxifying the addict and then teaching him or her how to live life without drugs. Most chemical-dependency facilities base their programs on the 12-step recovery philosophy developed by Alcoholics Anonymous.

Addicts reside in chemical-dependency treatment facilities such as the one Barry went to for an average of 2 to 4 weeks. (There are also outpatient versions of chemical dependency treatment.) On entry, an addict is given a thorough assessment to identify his or her individual treatment needs, and a treatment team then develops a plan based on these needs. The plan is two-fold: one regimen is for the patient's stay at the center, and a second regimen is for continuing care after the patient leaves the center.

A typical treatment team can include specialists such as a physician, clinical psychologist, pastor, nurse, dietitian, and counselor. By the time patients are ready to leave the center, they have usually reached the second or third step of the 12-step recovery process.

Returning home does not mean the patient has completed treatment; he or she must work through the rest of the 12 steps in evening meetings held at the treatment center or attend other 12-step meetings located throughout the community to complete the 12 steps.

Barry's Family as Part of the Treatment Program

The addict is not the only person who must change his or her behavior. To a lesser extent, so must the addict's family members. Therefore, most of the better treatment programs enhance the chance for successful recovery by offering programs for family members as well as addicts. While Barry was undergoing inpatient treatment, Richard, Cynthia, and Lisa attended intensive sessions designed specifically for addicts' family members. There, they learned to identify and change the enabling roles they had unconsciously played as Barry's drug use progressed to addiction. Lectures and films taught them about addiction, helped them understand how Barry became addicted, and taught them techniques to help Barry sustain recovery after he leaves the center.

Barry's family also participated in group therapy sessions with family members of other patients. Group therapy helps family members support each other through the painful process of dealing with addiction. It also prepares them for the liberating process of discovering how much better life is once the burden of addiction has been lifted from all concerned and the problems that underlie it have been resolved.

Alcoholics Anonymous and Other 12-Step Programs

Alcoholics Anonymous (AA), one of the earliest forms of treatment to emerge in the United States, created the 12-step philosophy. Remarkably, AA meetings are completely free of charge. Founded in 1935 by two doctors who suffered from alcoholism, the organization describes itself as "a fellowship of men and women who share their experience, strength, and hope with each other that they may solve their common problems and help others to recover from alcoholism."

Some 2 million men and women are members of AA. The organization believes that alcoholism is a progressive illness that cannot be cured, but can be arrested through total abstinence from any form of alcohol. The ultimate goal of this kind of treatment is to move addicts from uncontrolled alcohol use to no alcohol use. AA has helped millions of alcohol addicts give up alcohol and return to productive living. In some cases, alcoholics participate in AA exclusively and receive no other form of treatment. In other cases, AA serves as the aftercare component of short-term inpatient or outpatient treatment.

AA's 12-step approach has helped so many people recover from alcoholism that it has been adapted to help people recover from addictions to other drugs as well. There are now thousands of 12-step groups, and they cover a number of other drugs, such as Cocaine Anonymous, Narcotics Anonymous, Pot Smokers Anonymous, and others.

The circumstances that surround a person's addiction play a critical role in determining treatment needs. If the addict has a large array of personal and social resources available, such as a supportive family and a career to return to or pursue, the task of treatment is *rehabilitation* so that the addict can return to a productive life. If, on the other hand, the addict has few available resources and addiction prevented the acquisition of productive skills important to successful living, then the task of treatment is *habilitation*. After detoxification, the addict not only needs therapy to help replace maladaptive and self-destructive behaviors with more productive ones, but also must acquire basic skills needed to work and live successfully. Addicts lacking such skills must complete their education, develop social skills, learn job skills, and prepare themselves to make it in the real world.

Therapeutic Communities

Short-term treatment, such as that provided by chemical-dependency centers and similar forms of inpatient or outpatient treatment, generally works best for adolescents and adults who have strong family and social support, adequate social and job skills, and much to lose if they do not recover from their addictions. Adolescents and adults who do not have these support systems and who lack such skills require longer stays in treatment so that they have the time to acquire the skills needed to deal successfully with the world. Residential therapeutic communities meet this need.

The therapeutic community movement developed in the United States in the 1960s at a time when addiction, particularly heroin addiction, was thought to be untreatable. Like chemical-dependency centers, the 12-step programs, and similar forms of treatment, therapeutic communities also offer drug-free treatment. They address and reshape the dysfunctional anti-social, self-destructive behavioral patterns of addicts who have no family or no connection to their families and few life skills. The length of stay in a therapeutic community averages from 18 to 24 months and involves several stages.

Stage 1. Physical and psychological assessment.

Stage 2. Gradual introduction into the therapeutic community for intensive group therapy and continuous education. Therapeutic communities pioneered the idea of requiring patients to do the work of running the community as part of their treatment. As a result, patients develop a work ethic and confidence in their ability to take care of others as well as themselves.

Stage 3. Preparation for eventual reentry into the larger world. Therapeutic communities teach patients social and job skills as well.

Stage 4. Gradual reentry into the real world. Patients seek jobs and actually begin work while they still live at the community. The community supports them until they are ready to move out and assume full responsibility for themselves and the consequences of their actions.

Medications to Assist Treatment

Once people understand that drug addiction is a brain disorder, they inevitably come up with an obvious question. Why aren't we treating the brain? After all, addiction has parallels with other diseases that do use both behavioral alterations and medications. The successful treatment for atherosclerosis, for example, includes changes in lifestyle, but these are commonly aided and reinforced by medications that lower blood cholesterol levels. This model of treatment, combining medication with changes in behavior, makes perfect sense for drug addiction as well. Unfortunately, the answer to why the brain isn't treated is simple. With a few exceptions, we don't yet know how. This is the problem many neuroscientists are working on right now. They are trying to understand precisely how addictive drugs act on the brain and what kinds of changes these drugs produce after long-term use. Once scientists understand what kinds of changes have occurred, they may be able to develop medications that can return brain chemistry toward normal.

Having drugs that would either alleviate withdrawal or decrease craving would confer important advantages in helping addicts recover by putting them in a position to take advantage of the psychosocial treatment they do get. This is important because learning new behaviors through psychosocial treatment is crucial if the addict is to stay clean. Thus, with effective medications, the treatment for addiction could be just like the treatment for a number of other chronic diseases.

Like atherosclerosis, diabetes and hypertension are two diseases that require the combination of an effective medication and a change in lifestyle. Early in the progression of these diseases, lifestyle changes may be sufficient to reverse their course. Once significant physical changes (like fatty plaques in arteries) have occurred in the body, however, a medication is often required to stabilize the patient, but unless alterations in diet and exercise habits are made, the disease can continue to progress despite the medication.

Henry's Intervention

"I'm coming. I'm coming," said Henry's wife Gwen as she unlocked the door to their apartment, put down her groceries, and ran across the room to answer the persistent phone. "Hello?"

"It's me, Gwen. I'm in bad trouble. Baby, you got to come bail me out," Henry told his wife, his voice barely audible.

"Bail you out? From what," Gwen shrieked, both alarmed and angry.

"From jail," he said. "I sold a little heroin to some guy who turned out to be an undercover cop. He busted me."

"Henry! You *promised* me you'd never sell drugs. How could you," Gwen yelled at him.

"I know, baby. I know. But I owe Jake so much money — he made me start selling to pay him back. He was going to send his posse after me. I had no choice. They would have killed me."

"Maybe that would have been better," Gwen said acidly. She was furious with her husband and could barely contain her rage. "Where are you," she asked.

"Fifth precinct, corner of 17th and Elm."

"And how, exactly, do I bail you out? Where do I get the money?"

"Come on down to the station," he said. "They have people down here who will help us, tell us what to do."

"I'll get you out of jail, Henry. But this is it. I've had it," Gwen said, slamming down the phone and bursting into tears, completely frustrated.

What were they going to do? Henry hadn't had a steady job for more than 2 years. Their savings account was empty, and debts were piling up. Although Gwen had a good job, she just couldn't carry the load by herself. After his overdose, Henry had promised Gwen he would get off heroin and get a job. But he had not only broken his promise, he had dug himself in deeper. It had to stop. She just didn't know how to make that happen.

Eventually, Gwen's crying gave way to weariness. She pulled herself together, got her coat and purse, and left for the Fifth Precinct, wondering how on earth they were going to end this nightmare.

Gwen's worries ended when the judge thankfully intervened. Henry hadn't exactly been honest with his wife (addicts can be depended on not to be completely honest, for they are, after all, trying to minimize and conceal their self-destructive behavior, even from themselves). He'd been selling drugs for considerably longer than he admitted to Gwen, long enough and often enough to finally get arrested for selling drugs.

Henry's Treatment: Methadone

Soon after Gwen bailed him out, Henry had to go to court. His case was assigned to a judge who often heard drug cases and knew something about addiction. The judge had worked out agreements with local treatment facilities to provide assessments of drug offenders who came before him. Henry's assessment was positive for addiction, and the specialists who conducted it recommended that the judge place him in a methadone maintenance program. Henry admitted he broke the law and pleaded guilty to selling heroin. At his hearing, the judge gave him a choice: go to treatment or go to jail. Henry gladly chose treatment, grateful not to have to do jail time. The judge made it clear to Henry that if he failed at treatment, a jail cell would be waiting for him.

Methadone Maintenance

Neuroscientists are trying to develop a number of medications to help people recover from a variety of addictions. But there is one drug that has had clear success in treating addiction: methadone. It is used to treat heroin addiction. Originally developed as a long-acting synthetic opiate analgesic, methadone is now widely used to prevent withdrawal and craving in heroin addicts.

Advantages   Like heroin, methadone is an opiate agonist. But it binds to opiate receptors even more avidly than morphine does (remember, heroin is broken down into morphine in the brain). Therefore, it prevents withdrawal symptoms and cravings of addicts who have gone off heroin. Moreover, because methadone occupies opiate receptors, if the recovering addict did take heroin while using an adequate dose of methadone, the effects of heroin would be blocked. There would be no rush from shooting up.

Methadone is used as part of a detoxification program and as a maintenance drug. Because it produces little or no euphoria, most addicts who take methadone are able to work productively and lead normal lives, once they have successfully engaged in psychosocial therapy and have been rehabilitated. In some cases, methadone is used as a permanent substitute for heroin in addicts who repeatedly relapse into heroin use after being withdrawn from methadone. For addicts like these, long-term maintenance on methadone can be an effective treatment, especially when the alternative is a lifetime of heroin addiction.

Methadone has helped thousands of heroin addicts give up heroin, and most eventually get off methadone as well. The majority of addicts stay on methadone for 1 to 3 years. A minority take the drug for 10, 15, or even 20 years. Fewer still take it all their lives. Because methadone is administered orally, it greatly reduces the risk of becoming infected with HIV/AIDS, which is easily acquired through intravenous drug administration.

Disadvantages   Methadone has several disadvantages that have made using it for maintenance controversial.

1. Methadone can be and is abused by heroin addicts. Addicts who are in treatment may try to sell their methadone instead of taking it. This diversion into illegal markets can be a serious problem.

2. Methadone must be taken every day. This can be an advantage early in treatment because it brings the addict to the treatment center and into contact with treatment providers on a regular basis. This same characteristic becomes a disadvantage later, because it forces a recovering addict to come to the treatment center every day whether he needs to or not. This is inconvenient for those who are succeeding in treatment, and it hinders their ability to do many things that the rest of us take for granted, like go away for a few days.

3. Methadone is an opiate, and even though those who use it successfully no longer behave like addicts, they remain physically dependent. Some people feel that the only acceptable goal of treatment is a completely drug-free state, and they do not believe that physical dependence on methadone is an acceptable long-term treatment option.

4. Finally, methadone works only as a treatment for heroin addiction. Many addicts being maintained on methadone still use and are addicted to other drugs, such as alcohol (up to 25%) and cocaine (up to 60%).

The controversies surrounding methadone have led to some self-defeating practices. Fear of diversion has led some treatment programs to give inadequate doses of methadone to patients, who then relapse back to heroin use because they discover they can get high by "shooting over" the methadone (injecting heroin while taking methadone). Other programs, which fear the continued physical dependence inherent in methadone maintenance, set rigid limits on the length of time methadone can be used. Patients are withdrawn from methadone, no matter how they have progressed in other phases of their treatment. Still other programs, which fail to understand the need for comprehensive treatment, including psychosocial counseling, fail to provide counseling services. They either provide no therapy at all, or they provide inadequate amounts of therapy.

The failure to understand the neurobiology of addiction, and therefore the mechanism of methadone's therapeutic actions, has led some people to believe incorrectly that methadone is a substitute for all drugs of abuse. They advocate distributing methadone to all drug addicts as an inexpensive, cost-effective, efficient way to treat addiction. Such advocates completely misunderstand the mechanisms of addiction and the requirements of effective treatment. The end result of all these approaches is the same — failed treatment.

The high rates of treatment failure in methadone programs also have confused the way people view methadone treatment. People think it doesn't work because they don't know about the crucial differences in the ways treatment is delivered. Just as we wouldn't be surprised if an antibiotic failed to work at too low a dose or if it were given for too short a period of time, neither should we be surprised that poorly conceived or poorly delivered drug abuse treatment fails. These kinds of failures, however, should not be used to condemn the entire field. Research clearly shows that nationally designed methadone maintenance programs, which administer adequate doses of methadone and include appropriate psychosocial counseling, can effectively treat heroin addiction.

Others Drugs That Help Addicts Recover

Methadone is not the only drug that helps addicts recover. The following are also helpful:

1. **LAAM** — Because scientists understand the strengths and limitations of methadone for treating drug addiction, they continue to search for similar, but more effective medications. One that has recently received the approval of the Food and Drug Administration (FDA) is LAAM (L-alpha-acetylmethadol), an opiate agonist whose effects last three times longer than those of methadone. This means fewer trips to the treatment center for the addict and less diversion of the drug into the illicit market.

2. **Naltrexone** — This opiate antagonist has been available for a number of years for the treatment of heroin addiction. Because naltrexone is an antagonist, it blocks heroin's effects at opiate receptors. It can therefore be used only by detoxified addicts because it would induce withdrawal if given to someone who was still physically dependent on heroin (or methadone). It appears to be most useful in addicts who are highly motivated to recover. It has not gained widespread acceptance in treating heroin addiction, because most addicts don't want to take it. They prefer to

preserve the ability to get high. Naltrexone was recently shown to have beneficial effects in treating alcohol addiction, but it is too soon to know what its impact will be on the treatment of alcohol addiction.

3. **Buprenorphine** — A third class of drugs, with mixed agonist and antagonist properties, is also under study. One of these drugs, buprenorphine, has shown promise in treating heroin addicts. Its effects, which can last up to 3 days, are quite fascinating. It exerts agonist properties at lower doses and antagonist properties at higher doses. At low doses, it prevents withdrawal and craving. But at higher doses, its antagonist properties predominate and it can even precipitate withdrawal. It also shows antagonist actions when taken by someone who has just taken heroin.

   Because of its antagonist actions, buprenorphine is very safe. There is little chance of overdose. And it has a reduced risk of diversion to illegal markets because it can precipitate withdrawal in active addicts. Buprenorphine could therefore be a safe, long-lasting agonist with little risk of diversion. Active research is being carried out on this drug for treatment of heroin addiction.

4. **Clonidine** — Clonidine has no opiate-like properties at all. Instead, it inhibits the neurons of the locus coeruleus, which become overactive during withdrawal and are therefore responsible for many of the withdrawal symptoms. Clonidine has no abuse liability and is commonly used to help addicts get through opiate withdrawal with a reduced level of discomfort.

Sybil's Intervention

Sometimes doctors are the people who intervene in addiction and get the addict into treatment, as we see with Sybil.

What Sybil's CT Scan Revealed

Sybil — always in charge, always in control, the person everyone else turned to for advice and counsel — was terrified. Dr. Schwartz turned her world upside-down yesterday and now she had to face the hospital, a CT scan, and maybe a diagnosis of lung cancer.

As she walked toward the hospital entrance, Sybil was stunned to see patients dressed in robes and slippers standing outside smoking cigarettes beneath signs proclaiming that the hospital is a smoke-free facility. Several patients were tethered to an intravenous bag hanging from a rolling apparatus they dragged along behind them. These people have just undergone radical forms of cancer surgery, Sybil thought to herself, and still they smoke. She resolved on the spot to stop smoking, no matter how her CT scan turned out. She went inside and checked in, and an aide led her to the lab for the procedure. . . .

"There's no cancer inside your lungs, Sybil," Dr. McCloud began as soon as he introduced himself. "But this is a wake-up call. You must quit smoking, or you'll be back here in a year or two to let us remove a lung. Now give me your cigarettes," he commanded.

Shocked, Sybil meekly handed them over. Dr. McCloud crumpled them up, threw them in the waste basket, and gave her a prescription.

"Get this filled before you do anything else. It's a nicotine patch. Put it on this morning. It will help. I want to see you in 1 month. I don't want to hear that you have smoked a single cigarette. No excuses. I won't tell you your lungs look great. But you can probably avoid cancer if you quit now. Do it."

Sybil's Treatment

The Nicotine Patch

Sybil has found her way to treatment because her addiction had begun to threaten her health, and her doctor intervened. Doctors often intervene with smokers who have contracted life-threatening lung, heart, or other diseases as a result of their addiction to nicotine and who must quit smoking if they want to survive.

Sybil left the hospital and headed off for work, stopping at her pharmacy to fill Dr. McCloud's prescription for a nicotine patch and putting it on before she went to her office. That was just the very first step of her treatment.

FACTORS THAT COMPLICATE TREATMENT AND RECOVERY

For a variety of reasons, many people believe that treatment doesn't work. While this is simply not true, the myth persists for two major reasons. (1) No single treatment approach works for everyone. Different addicts need different kinds of treatment, depending on what drug or drugs they are addicted to and their individual histories and circumstances. (2) Treatment providers have not yet discovered a reliable way to determine in advance which kind of treatment will be best for which addict. Rather, they are often forced to engage in a trial-and-error approach until the right treatment can be found for a particular addict — that is, if the resources exist for several trials of treatment.

The Reality of Relapse

Another reason why people believe treatment doesn't work is based on misperceptions about how the process of recovery usually works and what the phenomena of craving and relapse mean within the recovery process. If you've ever tried to quit smoking cigarettes or you know someone who has, you almost certainly are familiar with relapse and the difficult path that recovery can take.

In fact, because of what we learned about recovery from addiction, we can now add one more facet to our definition of addiction. Drug addiction is a *chronic, relapsing* disease. It shares these characteristics with other diseases such as high blood pressure, adult-onset diabetes, and atherosclerosis. Atherosclerosis, for example, typically begins with voluntary behaviors, such as eating foods high in fat and not getting enough exercise. Drug addiction also begins as a voluntary behavior — the abuse of drugs. Just as high-fat diets and lack of exercise produce long-term changes in the arteries (such as the buildup of fatty deposits) that can result in atherosclerosis, drug abuse produces long-term changes in the brain that can result in addiction. People with atherosclerosis often "relapse" into their self-destructive lifestyles after a period of changed eating and exercise habits. Addicts are the same: they often relapse into drug-taking behavior after beginning treatment programs.

Viewed in its most narrow perspective, relapse is indeed a failure on the road to recovery. However, in the larger context of how people actually deal with many types of chronic diseases, *relapse can be viewed as part of the process that addicts go through as they strive to recover.* Many people are frustrated and even angry with addicts when they relapse, although they rarely experience the same emotions when a person with heart disease resumes a careless or abusive lifestyle and has a second heart attack.

The plain fact is that a repeating sequence of compliance with treatment and then relapse into self-destructive behavior is fairly common in the treatment of most chronic diseases. It takes time for people to accept that they have to change their lifestyles. Most people only gradually accept the changes that eventually allow them to survive.

About 30% of patients being treated for hypertension regularly take the medications their doctors prescribe, and about 30% comply with dietary restrictions designed to help them. As a result, some 40% of patients with hypertension have to be treated in the emergency room for episodes of acute, extreme high blood pressure. Similarly, only about 30% of adults who suffer from asthma take their medicine as prescribed. It is not surprising that many of the rest of these patients show up in the emergency room for treatment of acute asthma attacks. About 50% of patients with diabetes comply with routine medication. Less than 30% comply with proper diet and foot care. Consequently, about 50% of diabetes patients must be retreated within 1 year of diagnosis and initial treatment.

The bottom line is that it is hard for people to change their behavior, even if their very lives obviously depend on it. This is a characteristic that addicts share with the rest of us. Therefore, relapse into drug use, even after what seems to be an initially successful treatment, shouldn't surprise us.

CRAVING ELICITS RELAPSE

There is one special aspect of addiction that makes relapse even more likely to occur than with other diseases, and that is craving. Indeed, craving makes it common for drug addicts to relapse at least a few times before they settle into a long-term period of abstinence. Craving most commonly results from two causes. (1) The first has to do with the accumulation of maladaptive behaviors the addict learned as addiction developed. Recovery is a difficult process, and former addicts who encounter difficult times find it hard not to fall back into their old drug-taking behavior to solve life's problems. (2) The second cause is more subtle: *classical conditioning*. Addicts have deeply ingrained, conditioned responses resulting from thousands of training episodes when drug use was paired with some other stimulus. The longer an addict uses drugs, the more such associations are formed and the stronger they get.

Each conditioned response, or cue, has the capacity to elicit craving for drugs every time the recovering addict encounters it. Every day, the addict may encounter many cues to light up, inject, snort, chew, drink, or inhale the drug he or she is trying to give up. Each one of these cues initiates both a conscious awareness of the desire to use drugs and measurable physiological responses in the brain and body that mirror the craving and add to the addict's discomfort.

Let's look at the cravings of Sybil, Henry, and Barry: Even though Sybil is wearing the nicotine patch so she won't go into nicotine withdrawal, things get difficult once she starts her workday. For example, every time she picks up the phone, she unconsciously reaches for a cigarette. Henry's drug dealer, Jake, doesn't see things the way Gwen and the judge do. Jake keeps calling Henry, pressuring him to sell heroin again. Just the sound of Jake's voice makes Henry crave heroin. For Barry, seeing a magazine ad for rolling papers is all it takes to trigger a craving for a marijuana joint. It doesn't matter whether the addict is on some form of maintenance therapy, like Sybil, or has been drug-free for weeks, months, or, sometimes, years. Classically conditioned cues can elicit craving because the response to those cues has been well trained into the addict through many, many episodes of drug use.

Deconditioning

Addicts don't understand where their cravings come from (classical conditioning is unconscious learning), but all are likely to relapse if they don't learn how to recognize and work to neutralize these cravings. Some clinical researchers have been trying to "decondition" addicts by helping them identify and deal with the cues, or triggers, they unconsciously develop while they are addicted. Within the safety of the treatment center, these treatment providers expose and reexpose addicts to all of the triggers they identify until they no longer display an intense physiological response to them. These researchers are helping addicts extinguish the associations they learned through classical conditioning between ordinary things and drug-taking.

Some programs that help people quit smoking have been using a variation of this deconditioning technique for some time. Before smokers actually stop smoking, they are instructed to keep a diary and list all the things they may be doing every time they light up a cigarette. This list identifies their triggers. Once smokers recognize all their triggers, they are better able to withstand the craving that those triggers initiate.

An instructor who has taught smoking cessation classes for several years quit smoking herself using the technique of deconditioning. However, she recalls that one of her triggers eluded her and was not entered in her diary. About a year after she quit smoking, she took an out-of-town trip. After checking into her hotel and going to her room, she was overwhelmed by an intense craving for a cigarette. She realized with a start where it came from. One of the first things she had always done after going to a hotel room was to light up a cigarette. Just walking into a hotel room was a trigger for this woman. Because she hadn't neutralized it by identifying it before she quit smoking, it still had the power to make her crave a cigarette, even after a year of not smoking.

Deconditioning programs are not always completely successful. The conditioned responses are deeply ingrained, often from thousands of repetitions. So, in many cases, they can't be extinguished or "unlearned." Instead, the addict must learn new behaviors to use in place of the drug-using behaviors that create so much trouble.

So, a complex brain disorder with complex behavioral manifestations arising from unconscious, but powerful, forms of learning, creates a difficult problem for treatment providers. Even when treatment appears to be progressing well, periods of abstinence may still be punctuated by periods of relapse. With luck and persistence, the periods of abstinence become predominant, and the periods of relapse become minimal and eventually nonexistent.

CAN ADDICTION BE CURED?

Whether addiction can actually be "cured" is an open question. Twelve-step programs like AA are based on the premise that the changes induced by addiction are permanent and that addiction cannot be cured. That is why reexposure to the drug will lead to relapse into addiction. Even though proponents of the AA point of view do not believe that addiction can be cured, they do believe it can be treated successfully. Members are taught to *avoid relapse by avoiding drugs*. This is clearly the safest way to deal with addiction.

Some believe that moderate drug use may be possible for recovered addicts, suggesting that the changes caused by addiction are not permanent. But, while there is some evidence to support this point of view, it is not compelling. One possibility is that a subclass of addicts, perhaps those with relatively mild and brief histories of addiction, may be able to return to moderate drug use. Considering the downside risk of this approach, few treatment experts are willing to try or suggest it. The neurobiological evidence is not sufficient to support either point of view.

Changes in dopamine receptors seen in cocaine addicts appear to be very long-lasting. Changes in brain metabolism appear to endure as well. But it is difficult to interpret data like these because we do not know what role, if any, these changes play in the addictive process. So, from a neurobiological perspective, the question must remain open for now.

CHAPTER 11

THERAPEUTIC USE OF ADDICTIVE DRUGS: HOW CAN A BAD DRUG BE GOOD?

*A*lcohol, cannabis, opium, and other drugs have been used for thousands of years. People have long regarded these drugs with awe, taking advantage of them to treat disease. And, at the same time, they have degraded themselves by abusing them. Indeed, the abuse of drugs has caused tremendous problems for individuals, their families, and society. Consequently, the public has developed an understandable fear and loathing of drugs.

ONE REASON DRUGS HAVE BEEN PRESENT THROUGHOUT HISTORY

We tend to lose sight of the reasons why drugs have been with us throughout recorded history. Our ancestors gathered certain flowers, leaves, and bark, cultivated various plants, and fermented fruits and grains because they needed the yield from these plant products to treat disease, relieve suffering, and enhance religious and social rituals. For almost all of history, most of the world had no effective drugs to treat any disease. Alcohol, cannabis, and opium were often the only drugs available to treat injuries and illnesses, no matter how badly these drugs worked.

Our view of the history of drugs is obscured by the immense pharmacopoeia from which modern doctors and patients can select medicines. Most of us take these gifts of the modern pharmacy for granted. We simply cannot conceive of life without them. But consider this: Until the middle of the 19th century, alcohol was the only drug that existed to anesthetize patients for surgery. Imagine what it must have been like to have a tooth extracted without procaine (Novocaine) for the procedure and without codeine, or even aspirin, to relieve pain afterward. Even worse, imagine the amputation of a

limb without any anesthesia at all. We are so accustomed to being unconscious in the operating room that it is almost impossible to conceive of what
it must feel like to be fully conscious while a surgeon cuts us.

Furthermore, our ability to prevent and cure disease, routinely transplant
organs, and even replace limbs makes it equally difficult to realize just how
new modern medicine really is. From an historical perspective, it's brand
new. For thousands of years, mankind did without the twin benefits of effective medical know-how and medicines that actually worked. Barbers were
surgeons. "Medicine men" were pharmacists and physicians. Removing
blood with leeches was considered an effective treatment for people already
severely weakened by disease.

THE DEVELOPMENT OF MODERN MEDICINE

Even at the beginning of the 20th century, average life spans were almost
three decades shorter than life spans of today. Infant mortality rates were
heartbreakingly higher. Only a few generations ago, parents routinely named
several sons after their father in the hope that at least one son would survive
to adulthood to pass the father's name on to the next generation. Improve-
ments in public health and nutrition have made big differences, but it was the
development of vaccines and the introduction of effective antibiotics during
the middle third of this century that reduced infant deaths and lengthened life
spans. Infections that had commonly been lethal in the early part of the century were tamed. Now, they are easily treated and have become little more
than mere annoyances.

Surgery started to become an option in the 1840s, when a Boston dentist
and medical student named William Morton first demonstrated that patients
could be anesthetized for surgery with ether. Before that, the few surgical
procedures that did take place were done in response to life-threatening
emergencies — amputating an arm to treat gangrene, or a leg to treat an open
fracture. The only way to prepare a patient for these surgeries was to give
alcohol, and sometimes, if available, opium or cannabis by mouth (hypodermic needles had not yet been introduced) and to use various forms of
physical restraint. Doctors sometimes knocked people unconscious by
means of a blow to the head or by partial strangulation. More often, several
large men with strong arms held the patient down by force while the surgeon
did his work. Without a way to prevent pain, surgery simply wasn't considered, except in the most desperate of circumstances. Without the opportunity
to have surgery, many people died of the problems and diseases for which
treatment seems routine to us today.

From this perspective, it is easier to understand why our forebears regarded drugs such as opium, alcohol, and cannabis as crucially important medications. They were among the few drugs at their disposal. Even today, opiates have retained their place as important medications. Alcohol, by contrast, has lost its place entirely, and there is a vigorous debate about the value of cannabis.

Despite advances in almost every kind of modern pharmaceutical, opiates are still the most potent analgesics (painkillers) available. In fact, morphine sets the standard against which all newly developed pain medications are measured. On the other hand, alcohol has been replaced as an anesthetic by far more effective drugs developed over the century and a half since the discovery of ether's anesthetic properties.

Marijuana falls somewhere in between. The pharmacologic properties of tetrahydrocannabinol (THC), the active ingredient in marijuana, indicate that it might be a useful medication for more than just the two conditions it has been approved to treat. In addition, preliminary research indicates that other marijuana constituents may become effective medicines for a variety of other conditions. But, because of the public's widespread concern about marijuana abuse among young people, few pharmaceutical companies have been willing to study these compounds to the extent required to obtain marketing approval according to federal regulations. By the mid-1980s, only one company had obtained approval for a synthetic oral version of THC, called dronabinol (trade name Marinol), which is available by prescription for treatment of the nausea that often results from cancer chemotherapy and treatment of the AIDS wasting syndrome. The usefulness of THC and of other marijuana compounds for many other conditions remains to be determined. Any possible medical role of smoked marijuana is even less clear.

HOW CAN A DRUG THAT PRODUCES ADDICTION BE A MEDICINE?

Two issues confound comprehension when we try to understand the role of addictive drugs in medicine. First, how can a drug that causes the devastating behavioral changes of abuse and addiction be useful in medicine? This question is often phrased as: "How can a bad drug be good?" It can be rephrased more usefully as: "How can we take advantage of a drug's medicinal benefits and at the same time protect ourselves from its threats?" Second, how do we determine which drugs have medical application and which do not, especially when a drug has a long history as a folk remedy? When anecdotal stories are told about how some persons appear to have benefited from a drug, how can we show if that effect is real and that the population as a whole would benefit from it? If the evidence for effectiveness is just not clear, what do we do? Why do people get so upset about these issues, especially when a drug with abuse potential is involved?

To help answer the questions about which addictive drugs have medical benefits, let's briefly review the history of opiate use in this country.

HISTORY OF OPIATE USE

Opiates are among the most addictive drugs known to man; yet their medical usefulness has nonetheless spanned the millennia from ancient remedy to modern pharmaceutical. In the early 1800s, significant advances in chemistry and technology allowed pharmacology and medicine to leave the Dark Ages, so to speak, and begin the development of the modern drugs and medical procedures we make use of today. Advances in chemistry led the way, allowing scientists to break down the crude drugs found in nature (e.g., opium) into their specific components. New understanding about normal physiology and disease then launched a two-part quest that has driven the development of modern medical science and therapeutic drugs.

Scientists have led a persistent and continuous search for ways to isolate in the body the precise site of a specific health problem, and then have tried to develop a specific chemical compound (i.e., a drug) to treat that problem and no other. When scientists achieve this level of specificity, doctors can treat illnesses successfully, and patients are free from the side effects of less specific drugs. This ideal is hardly ever met, but we sometimes come pretty close.

SYNTHESIZING DRUGS FROM NATURE FOR GREATER EFFECTIVENESS AND FEWER SIDE EFFECTS

Scientists took one of the first steps in 1806, when they isolated from opium the compound morphine, named after Morpheus, the Greek god of dreams. The isolation of other opiate compounds soon followed. This ability to separate pure compounds from crude drugs signaled a major advance in the pursuit of ever more effective drugs. Throughout this early period, many important medicines were first used as crude plant extracts before scientists were able to isolate and then purify the compounds in those plants that have medically useful effects. In fact, scientists continue to search the tropical rain forests and ocean bottom for more of these amazing substances.

Once a compound has been isolated from a plant, medicinal chemists develop ways to synthesize that compound in the laboratory using pure chemical building blocks. The ability to synthesize compounds allows chemists to create medicines whose purity is guaranteed, whose potency is constant, and whose duration of action is predictable. For example, if your doctor prescribes a drug for you at a dosage of 2.5 mg, four times a day for 7 days, you can count on getting exactly the same amount of medication each time you swallow a pill (as long as you take the medicine as prescribed). Being able to tinker with chemicals in the laboratory also means that chemists can change the properties of those chemicals and perhaps increase the specificity (producing fewer side effects), potency, and duration of action (methadone acts about six times longer than heroin).

ADDICTION TAKES HOLD

An unrelated advance in the middle of the 19th century, the introduction of the hypodermic needle, allowed physicians to administer medicines directly into the bloodstream for faster, more effective action. This medical advance had the unintended side effect of accelerating and intensifying morphine abuse. Morphine was given to soldiers in the Civil War to relieve the pain caused by brutal injuries. So many soldiers became addicted that morphine addiction came to be called the *soldier's disease*. After the war, soldiers recruited others to use morphine. This appeared to contribute to the increase in opiate consumption throughout the rest of the century.

By 1900, opiates were found in many "medicines," although, in the absence of labeling requirements, people who took these drugs were unaware of their contents. These medications rarely had any therapeutic value, but some at least did make people feel a little better. It is not surprising that opiate abuse and addiction became menacing social problems.

THE SPREAD OF UNSAFE, INEFFECTIVE "MEDICINES"

At about the same time in the late 1800s, another trend was developing. People began to get angry about useless and harmful medications. Although it seems difficult to believe, drugs then were not regulated in any way. They did not have to be safe. They did not have to be effective. Anyone could manufacture a "medicine" (safe or toxic), make claims for its curative powers (true or false), and sell it to sick people desperate for relief. Because no one was required to list ingredients on labels, many medicines contained various forms of opiates, cannabis, cocaine, and alcohol, but people had no way of knowing that the medicines they bought contained these drugs. Thus, as the consumption of medicines containing these substances steadily increased, abuse and addiction to them spread, and public support for drug regulation intensified.

The public's desire for appropriate control and regulation of drugs, partly driven by pervasive drug abuse, was further intensified by the sale of many medicines that were completely worthless or dangerous or, at times, both.

In response to growing public concern about the prevalence of dangerous, worthless "medicines" and about the spread of addiction, Congress passed two entirely different sets of laws in the early part of the 20th century: one to regulate the sale of medicines and one to control drugs of abuse. The first set of laws was designed to govern the development, safety, and effectiveness of new medicines distributed to the public. The second set was intended to reduce or eliminate drug abuse, drug addiction, and drug trafficking.

FOOD AND DRUG LAWS

The first of the drug approval laws, the Federal Pure Food and Drug Act of 1906, addressed the purity of both food and drugs. Following a series of deaths from the distribution of a highly toxic preparation a few decades later, Congress amended the act in 1938 to add labeling and safety requirements for all newly developed medicines. This amendment also established the Food and Drug Administration (FDA), a government agency responsible for monitoring the enactment of these requirements.

Congress added an additional amendment in 1962 after the thalidomide disaster took place in Europe, where a large number of mothers who took the drug during pregnancy gave birth to babies without arms or legs. This 1962 amendment strengthened safety provisions and added the requirement that all drugs marketed to the public must be effective in treating the problem or disease that the manufacturer designed the drug to relieve or cure. Congress made this amendment retroactive to 1938 to include all drugs that had been introduced since that date. Strengthening safety provisions and adding this "efficacy" requirement (meaning that a drug must reliably produce its claimed effect) further protected the public from unsafe and ineffective drugs.

THE DRUG ABUSE CONTROL LAWS

The drug abuse control laws had a dual task: to reduce drug abuse and, at the same time, to preserve our ability to use even abused drugs in medicine when they are needed. Like the food and drug laws, Congress passed and amended the drug control laws throughout the 20th century. And, just as Congress established the FDA to oversee the food and drug laws, Congress established the Drug Enforcement Administration (DEA) to oversee the drug control laws.

To control drugs and at the same time make them available for use in medicine, these laws categorize drugs according to two primary criteria: their usefulness in medicine and their potential for abuse. The categories, called *schedules*, contain every class of drug that people abuse. Only drugs with no proven use in medicine, such as heroin, lysergic acid diethylamide (LSD), and marijuana, are placed in Schedule I, which prohibits producing, distributing, or using these drugs except in research. Researchers must apply for special licenses from DEA to obtain the Schedule I drug they want to study.

Other drugs, such as cocaine and morphine, which have great potential for abuse, but also clear medical uses, are in Schedule II. Their use is permitted in medicine, but is carefully restricted and monitored. As the perceived danger of abuse of medically useful drugs diminishes, they are placed in Schedule III, Schedule IV, or Schedule V, which means that even though these are drugs people might abuse, doctors can nonetheless prescribe them to patients for use in medicine.

Schedule I drugs are most tightly controlled; Schedule V drugs are least tightly controlled. Drugs with no abuse potential that have been approved for marketing by FDA are not controlled at all, and doctors may prescribe them without the kinds of restrictions that apply to abused drugs. Morphine and methadone, for example, are in Schedule II. Buprenorphine, which is not widely abused because it can have antagonist actions, is in Schedule V. Antibiotics, which have no abuse potential, are not in any schedule.

Although written separately to address distinct problems, both sets of laws end up sharing a common purpose. They allow us to have safe and effective medications, while protecting us from drugs that might be useless, dangerous, or addictive. There is a long list of nonaddictive drugs that never were, or are no longer, available for use as medicines because they were either not safe or not effective. Ineffective drugs have been a particular plague to patients with devastating or fatal diseases like cancer and AIDS. When people are desperate, even just for the *hope* of a cure, they are willing to try anything. Despite FDA protections, charlatans still sometimes take advantage of this situation, producing and selling a drug they know is worthless just to make money. Other times, patients might actually believe that they have found a useful treatment, even though research has not established safety and efficacy. (Many cancer patients' belief that Laetrile, an extract of apricot pits, could cure their cancer is a good example.) In either case, precious time, money, and emotional resources can be wasted because the alleged cure is no cure at all.

APPROVAL OF NEW DRUGS

If the issues of safety and efficacy are so difficult to deal with, just how does a drug become approved for medical use? In the United States, a drug is approved for marketing to the public only after it has passed a complex and systematic battery of safety and efficacy tests in a carefully controlled process monitored by the FDA. Similar procedures are in place throughout the rest of the developed world.

In the United States, only the FDA can approve a new drug for marketing. The process is so difficult that it can take up to 10 years. The paperwork for the typical application that pharmaceutical companies send to the FDA for new drug approval can fill more than a dozen large file cabinets. (Recently, however, several changes to FDA law have simplified and shortened this process.)

The new drug approval process starts in the chemistry laboratory and continues through an array of animal tests before the new drug may be given to any human. The animal tests help to determine whether a drug might be effective and whether it is reasonably safe. A preliminary application to the FDA describing the evidence from animal testing sets the stage for the first human studies. Even then, a new drug is typically given first to a small number of healthy human volunteers to establish that it is safe enough to use on sick patients. Only after the first period or trial of human testing (phase I) can the new drug be tested for its ability to treat the condition it was designed to treat. Sick human patients are then enrolled in research studies designed to determine whether the drug actually does what its manufacturer claims it can do. This period of testing is called phase II.

DOUBLE-BLIND STUDIES

The most critical tests of a drug's efficacy are carried out during phase II studies. They represent the gold standard for validating the effectiveness of a drug. These tests are called *double-blind trials*.

In a double-blind trial, patients are randomly assigned to either an experimental or a control group. Patients in the experimental group receive the experimental drug in question; those in the control group receive a standard treatment. If no treatment exists, patients may be given a placebo (i.e., an inactive substance) as long as it is considered safe to do so.

Protection of patients is of paramount concern in double-blind trials. Therefore, all patients must give informed consent, documenting that they understand the potential risks and benefits of the medications that will be given in the trial. In addition, treatment is never withheld if it is available. Placebos are used only when no treatment is known. Otherwise, the new drug is compared with existing drugs. A drug study is "blind" when the patients don't know whether they are getting the experimental drug or the standard treatment (or placebo). A study is double-blind when the doctors and nurses conducting the study also don't know who is getting the experimental drug.

Before a double-blind trial, the drug and the placebo are put into identical packages and coded so that no one can identify which is which. During the trial, patients' responses to what they are taking are carefully recorded. After the trial, the code is broken. Only then can the researchers determine whether positive results noted in patients were caused by the experimental drug. (In some drug studies, both the doctors and the patients know that an experimental drug is being used. This is called an *open-label study*.)

If a drug fails to demonstrate efficacy in a double-blind trial, it is nearly impossible to convince scientists and doctors that it is worth using. However, once a drug has demonstrated efficacy in a series of double-blind trials, it is gradually introduced to the public by being administered to progressively larger groups of patients. This is done in additional carefully monitored trials to ensure that unexpected side effects or previously undetected problems don't appear. These larger trials constitute phase III of the approval process. Still, even after approval for mass marketing has been granted, monitoring continues. Efficacy is not enough. Safety, too, is crucial.

RISK/BENEFIT ANALYSIS

The safety of a drug is a relative issue. Most cancer chemotherapies, for example, are extremely toxic. But the alternative is usually death, so a certain level of danger from the drug itself can be tolerated if its benefits are significant enough. No drug is completely safe.

For example, digitalis, used for many decades to treat heart failure, can be toxic at doses only slightly higher than the therapeutic dose. Despite its dangers, digitalis can be a lifesaving drug. It has been used because it is so effective in helping the heart beat more forcefully. Aspirin can cure your headache, but might upset your stomach. Most people are willing to take that small, known risk for the benefit of not having a headache. Each time we decide to use a medication, we are making (or at least accepting), either implicitly or explicitly, a risk/benefit analysis that determines whether it is worth it to us to take that medication. This is in essence what the FDA does during its new drug approval process.

Sometimes these risk/benefit analyses are easy; other times they are not. It probably makes sense to use a drug that is toxic if there is no alternative or if the disease for which the drug is being taken is severe. It doesn't make the same kind of sense to use a toxic drug to treat a mild or temporary problem. Most choices lie somewhere between these two obvious extremes.

SOME DRUGS OF ABUSE ARE EFFECTIVE MEDICINES

We can use analogous logic to analyze the medical use of drugs that can cause abuse or addiction. We want to have drugs such as morphine and cocaine available as medicines because we need them. We also need other stimulants, sedatives, inhalants, and hypnotics because they all have significant medical benefits. Moreover, all these drugs are effective and safe when used properly, for example:

- Opiates are irreplaceable as analgesics for moderate or severe pain.

- Cocaine has its special uses as a local anesthetic.

- Other stimulants are used to treat attention deficit disorder in children.

- Tranquilizers are used by many people to treat anxiety that is disrupting their lives, and some are useful in the treatment of epilepsy.

- Nitrous oxide is used in some dental procedures.

- Many people have relied on a hypnotic to get to sleep at times.

Deciding whether a drug with abuse potential should be made available as a medicine is a risk/benefit analysis, often a complex one. In some cases, as with opiates, both the benefits and the risks are clear and compelling. In other cases, as with marijuana, neither the risks nor the benefits are quite so clear nor nearly so compelling.

Whether difficult or not, we must make decisions about addictive drugs' risks and benefits. We must deal with them in all their complexity or risk two unfortunate, alternative outcomes. (1) If we overrespond to the dangers a drug presents, we may forfeit some medically important treatment capability. This is a terrible outcome for patients who may need that capability and should be unacceptable to the rest of us. (2) If we allow a drug to become too easily available, then many people will abuse it. This creates a different set of problems and is equally unacceptable. Experience has shown, however, that with proper policies and oversight, we can have the medical benefits these hazardous drugs promise while we protect ourselves from the risks they pose when people abuse them.

Opiates: Most Effective Pain Relievers Available

The struggle to achieve a balance with respect to opiates is set in the framework of our initial experience with opiate addiction in the last century. At that time, unrestrained use of opiates led to a tidal wave of opiate addiction. That wave of addiction produced a backlash against opiates. For most of this century, the pendulum of public opinion swung away from attitudes that permitted appropriate opiate use in medicine. People became so concerned about the addictive potential of strong opiates that they extended their concerns to all areas, including medicine. As a result, doctors were reluctant to administer appropriate doses of opiate analgesics, and medical use of these drugs was miserly. Opiates were administered in doses too small to be totally effective and on rigid schedules that ensured unnecessary suffering by allowing pain to return every few hours.

The therapeutic failure to appropriately relieve pain occurred even in patients who were not at risk for developing physical dependence and addiction simply because they did not take the medications long enough for these conditions to develop. But the problem was more severe when individuals needed to take opiates for longer periods of time to relieve pain. People were afraid that they or their loved ones would become addicted to opiates. They either refused to take opiates themselves or refused to allow doctors to give opiates to members of their families. Because so many people had concluded that opiates would cause the same havoc in the lives of pain patients as they do in the lives of addicts, doctors and nurses were too often afraid to administer opiates appropriately.

Over the years, as research with both pain patients and addicts has progressed, it has become clear that pain patients treated properly have a very low risk of becoming addicted. How can this be? The answer lies in the factors that determine the difference between physical dependence and addiction — and explains why we took so much time to clarify the differences between these two states in Chapters 8 and 9.

The Difference Between Pain Patients and Addicts

Pain patients who take opiates for long periods of time can become physically dependent on opiates, just as addicts do. They would experience withdrawal symptoms if they suddenly stop taking opiates, just as addicts would. But, unlike addicts, once pain patients stop taking opiates for good, they rarely continue to seek opiates afterwards. Addicts lose control of their drug-seeking and drug-taking; pain patients rarely do so. Once their pain is gone and they stop taking opiates, they resume their normal lives.

The Difference Between Physical Dependence and Addiction

The extreme differences in the lives of these two sets of people — those with physical dependence and those with addiction — who chronically self-administer opiates are truly remarkable. Both use the same kind of drug. Both self-administer their dosages. How can the outcome be so different? Two factors can explain this: the person's reason for using the drug and the person's desired outcome from using the drug. The addictive properties of the drug are the result of the interaction between our biology (in this case, our brain), the environment (in this case, the reason for taking the drug), and the drug (which will produce physical dependence under any circumstances, but addiction only when specific environmental factors are added). Considering any one of these three factors in isolation can create an incomplete picture and can lead to incorrect conclusions about the danger and medical usefulness of the drug.

In the past, all the focus had been on the drug. But the drug can produce addiction only when other influences come into play. *The most important influence is the use of drugs to produce pleasure or reduce stress* — that is, the abuse of the drug.

People who take opiates for pain over long periods of time develop tolerance and physical dependence just as addicts do, but their responses to these phenomena differ somewhat from those of addicts. Tolerance to the sedative effects of opiates and to their ability to disrupt thinking occurs before tolerance to their pain-reducing effects. Thus, when people first take morphine for pain, they become tired, have trouble concentrating, and just can't pay clear attention to the world. After a few days on the medication, however, people notice that these effects begin to trail off and can even disappear while pain relief is maintained.

Doctors have learned that pain patients can keep taking larger and larger doses of opiates for pain as they become more and more tolerant to their effects. In fact, when we are in pain, we appear to have an almost limitless ability to develop tolerance to opiates. For example, the standard analgesic dose of morphine is 10 milligrams (10 one/thousandths of a gram) every 4 hours, totaling up to 60 mg per day. Some cancer patients, however, can take many *grams* of morphine each day, enough to kill several healthy people. Therefore, when opiates are used properly in medicine, the only effect of the development of tolerance is that patients require more drug. If there are no serious side effects, drug tolerance is not a significant problem (although the importance of side effects should not be minimized).

Lack of Understanding Leads to Needless Suffering

Even though we have come a long way in understanding the difference between physical dependence and addiction, some people who need opiates, or some other potentially addictive drug to treat pain or other medical disorder, still find cultural, psychological, or legal obstacles in their way. Many patients who know or learn that there is a high likelihood of physical dependence with long-term drug use mistakenly believe that they will become addicts. Some parents have even prevented their severely ill children from using opiates because of this fear. This confusion comes about in large part because the difference between physical dependence and addiction is still not clearly understood.

All things considered, almost a century after we began to put our worst opiate abuse problems behind us, we have reached a more reasoned understanding about how these remarkable drugs can and should be used medically. First, scientists and physicians developed ways of using opiates that optimized their analgesic actions while minimizing their addictive risks. Even so, it took decades for attitudes to change. Advocates for appropriate pain treatment worked to do this during the 1970s and 1980s especially, before the medical profession and the general public were ready to accept what the pain treatment experts had learned in the laboratory, at the hospital bedside, and in the hospice.

Effective Pain Treatment Today

Generally, the distance we have traveled in understanding how to use opiates appropriately in medicine has been remarkable. Now, instead of treating surgery patients for pain with doses of morphine that are too small and too far apart, many hospitals use patient-controlled analgesia. With this method, the patient has a computer-controlled pump and administers as much morphine as is needed (the computer prevents the patient from taking too much) when it is needed. Some people feared that giving control of opiate administration to the patient would produce addicts. In reality, the opposite has happened. Patients who control their administration of opiates actually use less drug than those who must ask for their pain medicine from a nurse, they report better pain relief, and they rarely become addicted.

The fact that appropriate use of opiates does not lead to addiction has turned out to be among the most important lessons to be learned about treating pain properly. This understanding could emerge only after people realized that *physical dependence is not addiction.*

The science of addiction, which has now taught us that the concept of addiction is different from the physiological state of physical dependence, has something to teach us about pain treatment as well. What we have learned in studying addiction isn't just about addicts. There are medically sound, morally compassionate reasons to distinguish between physical dependence and addiction, because at least some drugs that people abuse and can become addicted to can also prevent suffering.

IS MARIJUANA MEDICINE?

Although we have reached a reasonable accommodation in balancing the risks and benefits of medically used opiates, we are far from such a reconciliation as we grope for a way to evaluate any potential medical use of marijuana. If our drug laws and policies can make room for the appropriate medical use of the highly addictive opiates, can they also make room for the appropriate medical use of the cannabinoids, which do not have the same addictive potential as opiates? Can we make a risk/benefit analysis with marijuana as we have with morphine? Can we find some reasonable middle ground between the outright prohibition represented by marijuana's current Schedule I classification and the unregulated availability promised by initiatives voters have passed in some states.

For starters, the risk/benefit analysis for marijuana isn't that simple. The debate over the medical use of marijuana in the United States is at best confusing and at worst so fraught with emotional arguments, impassioned but diametrically opposed beliefs, and frenetic political activity that it is almost impossible to sort out what is accurate and what is not. These are difficult circumstances under which to conduct any kind of analysis. Let's try to dissect this situation.

In some respects, the issues that surround the question of marijuana's potential medical use in the latter half of the 20th century are similar to, and in some cases parallel with, those that surrounded opiates in the latter half of the 19th century. Just as easy availability of opiates led to the widespread addiction that left a legacy of prohibiting opiate use for any reason, several factors led to an explosion of drug use and addiction in the 1960s and 1970s. This produced a similar legacy of concern about the use of marijuana for any reason. But the parallels are not precise.

The Drug Epidemic of the 1960s and 1970s

The explosion in drug use that began in the 1960s continued to accelerate and, for the first time in history, enveloped adolescents a decade later. By the end of the 1970s, one third of adolescents, two thirds of high school seniors, and nearly three fourths of young adults had used an illicit drug, mostly marijuana. Eleven percent of seniors used the drug daily, with devastating consequences in terms of addiction, their need for treatment, and their drug-induced inability to fulfill the necessary tasks of adolescence: to learn how to lead productive lives as adults.

A political movement evolved in the early 1970s to legalize drugs. This movement appeared to parallel the rise in drug use among adolescents throughout that decade. Faced with the public's antipathy to legalization in the 1980s, legalization proponents latched on to "medical marijuana" as a way to advance their cause. More recently, many cancer patients and patients suffering from AIDS wasting syndrome have obtained marijuana illegally. They have smoked it and found that it helped to allay their anxieties, reduce their nausea, or increase their appetite so they could resume eating. It is difficult for these patients, their families, friends, and advocates to understand why they cannot legally access something that helps them, when every teenager in the nation seems to know how and where to obtain marijuana just for the asking.

Finally, the wave of cocaine and crack addiction that followed the explosion of marijuana use in the 1960s and 1970s created a further revulsion to marijuana in the public's mind. Consequently, there was little or no public support for exploring any possible use that marijuana derivatives or the crude drug itself might have in medicine.

All these varied concerns, perceptions, and conflicting viewpoints about marijuana use make a complicated situation exceedingly complex. To make some sense out of this complexity, let's review several issues that may guide us to some reasonable resolution of this problem.

Not an Approved Medicine

The first issue concerning the use of marijuana in medicine is that neither smoked marijuana nor any marijuana constituent except THC has been approved as a safe or effective medicine. The drug approval laws demand that certain scientific standards of safety and efficacy be met before a drug may be legally marketed as a therapeutic agent. These laws are intimately related to the quest to improve medicine that we described previously. For almost two centuries, the evolution of drug therapy has always been from crude plant extract to pure compound. In every case in which it has been possible, medical scientists have isolated and synthesized pure compounds to ensure specificity of action, purity, and potency. And this is just what happened with marijuana.

Scientists first isolated and later synthesized several cannabinoids from the plant. As noted earlier, the FDA has approved one of those cannabinoids, THC, to relieve chemotherapy-induced nausea and to treat AIDS wasting syndrome. For reasons we will discuss in the following text, treatment of these two conditions constitutes the only two medically approved uses for THC.

Any good pharmaceutical scientist might ask why administer a crude extract (smoked marijuana) when we have the purified active ingredient. Because they have always held that the pure chemical is the ideal therapeutic agent, almost all medical scientists are skeptical about the motivation behind the outcry for the use of smoked marijuana rather than pure THC. To these scientists, it sounds like regression, not progression.

At the outset, the scientific community is lined up against the use of smoked marijuana as a therapeutic drug. Using it flies in the face of history and experience. The active ingredient is available, so what's all the uproar about? If meeting FDA requirements were the only issue, there would be no debate. The case is already closed. Smoked marijuana meets none of the FDA regulatory requirements for safety and efficacy, and no one seems willing to invest the money required to test it. In addition, because of the drug control laws, every pharmaceutical company knows that a product with abuse liability will come under special scrutiny from DEA. That diminishes profits. Is anyone surprised that we know so little about the potential therapeutic benefits of marijuana and its constituents?

Marijuana Abuse and Its Impact on Children — and on Research

The second issue concerning the use of marijuana in medicine is the public's concern about the impact of marijuana abuse on society, particularly on the young. Marijuana is a dangerous drug. We have known for decades that people use it compulsively and that it disrupts lives. Tens of thousands of young people end up in drug abuse treatment programs to deal with their involvement with marijuana. Recent evidence indicates pretty clearly that marijuana can cause physical dependence and that it acts on the same brain circuits as addictive drugs like cocaine and heroin. Marijuana's abuse potential must be weighed in any risk/benefit analysis. Denying that risk is being dishonest. The public's concerns about marijuana are based in reality.

Furthermore, the concerns about marijuana have had beneficial effects. For example, drug use by young people fell significantly during the entire decade of the 1980s. But there was another, not so beneficial, effect. In fact, the pharmaceutical industry cites "public concerns about marijuana" as one of the main reasons why we know so little about the potential therapeutic effects of either marijuana, THC, or any of the other cannabinoids the drug contains. Industry representatives ask why any company would want to work with a controversial product or try to develop its chemical derivatives, when it could spend its time and money more fruitfully on other products? If there were a more compelling health need to get into that fight, some companies might do it, but only one has so far.

Negative public attitudes have also prevented the federal government from investing in the study of medical uses of marijuana. Thus, all the pressure has been for research to *prevent* marijuana abuse and to treat its consequences once use has gotten out of control. By the 1980s, even as the federal health agencies were trying to convince physicians to use more opiates to treat pain, no official of the federal government would dare make any statement about marijuana short of condemning it outright. To do otherwise was to risk being labeled as "soft on drugs." The result, intended or not, is that research on the medical uses of marijuana in general and cannabinoids in particular simply didn't occur. Now that increasingly large numbers of desperately and terminally ill people are asking if marijuana will help them, we cannot give them the answers they need.

Compassion for the Dying

Add to this atmosphere the issue of compassion for those who are dying. Supporters of compassionate use believe that the need to treat devastating and fatal diseases should override concerns about safety and efficacy and that desperately sick people are entitled to any form of medical care, approved or unapproved, which may help them. They have a morally compelling point of view.

Parallel With Heroin

The debate about marijuana is basically a replay of the debate about the use of heroin to treat pain. Although heroin was introduced as a treatment for morphine addiction (how's that for irony!), its potent addictive potential long ago drove it from the legal market. Until the 1980s, heroin did have properties that no other opiate could equal. It was more potent than morphine, and it had a more rapid onset of action. In an environment where pain was treated only after it appeared, these were crucial advantages, especially for cancer patients in chronic pain. These patients were often physically wasted, so they had little muscle mass into which to inject a drug. A potent drug that acted rapidly and could be given in a small amount was a true therapeutic advantage. So, it was possible to argue that, for reasons of compassion, heroin should be made available to treat cancer pain. For health care providers who treat pain and for the scientists who study it, arguments for compassionate use were hard to dispute. But concerns about addiction were overriding, and heroin never became available to treat pain in this country, even though it is hard to understand why addiction should be a concern in the face of death. Now, there are drugs more potent than heroin, and the guiding principles of pain management compel treatment that prevents pain from coming back, so heroin's advantages are no longer important and the efforts to legalize heroin for pain treatment disappeared.

When people today apply the "compassionate use" argument to marijuana, we are forced to listen. True, marijuana is not pure. True, it has not met any FDA requirements. But it seems to work for some people, especially the severely ill. Why not let them use it? We can't really argue with that point of view.

Legalization Proponents' Agenda

Unfortunately, marijuana legalization proponents, who seek to overturn the drug *control* laws, make embracing the compassionate use argument more difficult than it should be. Their true goal is to allow anyone to use marijuana more or less the way alcohol is used now. If the arguments of legalization proponents become convincing to most Americans, we will simply set up a new wave of marijuana abuse. These proponents have thrown their support behind the supporters of compassionate use, betting that any approved use of marijuana will simply be a step on the road to complete legalization. With all these points of view, some above-board and some covert, it is no wonder why the risk/benefit analysis of marijuana use in medicine is so difficult to figure out.

Any risk/benefit analysis requires good information and the ability and desire to analyze the issues objectively. But, when all the factors are not readily apparent or available and are further obscured by political concerns, there seems little chance of coming to a rational decision about marijuana use any time soon. Instead of learning from history, we seem doomed to repeat it, unable to discover the proper place of marijuana in medicine without conducting large-scale social experiments, which will probably teach us just what we learned about opiates. This is too bad because the potential cost is high.

Resolving the Marijuana Dilemma

Can we help the dying and let science decide if marijuana has medical value at the same time? It's certainly possible. First, reasonably designed clinical trials should be run to determine whether dronabinol, other cannabinoids, or marijuana itself, have efficacy in any of the clinical conditions supporters claim. Because this is not only a public health issue, but a political one that is under intense public scrutiny and because no pharmaceutical company seems willing to accept the responsibility for exploring the issue, the National Institutes of Health should be enlisted either to oversee or run clinical trials and taxpayers should pay the bill. This has already begun to happen.

Clinical trials for marijuana would not be easy to design because it is almost impossible to "blind" a trial with a psychoactive drug. Patients figure out pretty quickly what substance they are taking. Nevertheless, appropriate trials can be designed and run. The public should be willing to support such research, because our experience with opiates has taught us that it is possible to have the medical benefits without sliding down the slippery slope to epidemic abuse.

The risks of marijuana abuse resulting from medical use are dwarfed by the risks from more standard causes, such as failure to work hard at drug education and prevention. States should wait for the research data before legalizing marijuana in ways that make it likely that it will be abused. The state ballot initiatives that legalize controlled drugs as medicine, such as those in California, Arizona, and other states are good examples of what not to do. It seems unwise to shift the responsibility for new drug approval from scientists and doctors to voters, but that is a decision voters will make in any case.

Finally, while research is being conducted to provide answers, the outcry to make marijuana available to patients who are terminally ill will continue. So why not try compassionate use? Make a special exception to permit physicians to prescribe medical-grade marijuana for severely ill patients while we wait for answers. Inform patients receiving the drugs and their families of the risks inherent in such an action. Make it clear that marijuana has not met the tests required to approve it as a medicine.

Such a step would allow us to try to relieve the very real suffering of people who are dying while we buy the time to obtain answers and, most likely, improved and more effective drugs. At the same time, it would allow us to protect citizens from the very real suffering that occurs as a result of drug abuse and addiction. It does not have to be an either/or choice. We need to remember what we have learned about opiates: that addictive drugs can be used in medicine without spreading addiction. With compassion for both the dying *and* the addicted, we can relieve suffering and protect people from addiction at the same time, if only we are wise enough to learn the lessons of history.

CHAPTER 12

SUMMING IT ALL UP

*I*n this book, we presented an array of people, some real, some invented, who have lost control of their behavior because they are addicted to drugs.

- Allison and her friends are dead because they got drunk and then drove while intoxicated.

- Sybil has damaged her lungs by smoking cigarettes and may still end up with lung cancer or emphysema.

- Chris is selling his uncle's possessions and embezzling funds at work to finance his cocaine addiction.

- Henry was arrested for selling heroin.

- Barry repeatedly stole money from his parents until a family intervention got him into treatment.

We don't know whether treatment for Sybil, Henry, or Barry will ultimately be successful, because relapse is a real risk for all of them. And we don't know if Chris will find his way to treatment at all.

As these addicts initiated and then continued their drug use, most underwent changes in behavior that eventually made them almost unrecognizable to the people who love them — and to themselves.

Throughout this book, we have tried to explain what underlies the behavioral changes seen in drug addicts. Let's briefly review what happens.

DIRECT EFFECTS OF THE DRUG

At least two sets of mechanisms lead to and maintain addiction. First, there are the *direct effects of the drug* itself. When the brain is repeatedly exposed to drugs, tolerance and physical dependence inevitably develop. Based on the difference between heroin addicts and pain patients who are taking strong opiates as medication, it is now reasonable to believe that additional changes take place in the brain that account for the psychological dependence that distinguishes addiction from tolerance and physical dependence. Neuroscientists have now described alterations in receptors, second messengers, enzymes that synthesize neurotransmitters, and other components of the brain reward system in the brains of human addicts and animals that have been chronically exposed to drugs. There is even evidence suggesting that chronic drug exposure might reduce the number of proteins that help support the structure of the dopamine-containing neurons in the ventral tegmental area, shrinking them and reducing the number of dendrites they contain.

The next few years will produce a startling array of information about what occurs in the addicted brain. Although the brains of addicts are different from normal brains, it is not yet clear whether the changes that have been described are related to tolerance, physical dependence, or addiction per se.

THE LEARNED BASIS OF ADDICTION

Even though exposure to drugs is necessary to establish the conditions that lead to addiction, a second set of mechanisms must also be activated to produce addiction. These are the mechanisms that underlie the different kinds of learning that take place as addiction develops: (1) operant conditioning, (2) classical conditioning, and (3) explicit learning and memory. Indeed, the role of learning in addiction is so important that one way to describe addiction is as the sum of the several kinds of learning that take place while the brain is adapting to the presence of a drug.

Operant Conditioning

Operant conditioning teaches drug users that taking a drug produces pleasure. As a result, drug users take more of the drug. Because drugs of abuse directly activate the brain reward system, the positively rewarding qualities of the drug powerfully reinforce drug-taking behavior. Operant conditioning also teaches drug users that a drug can relieve unpleasant feelings, such as anxiety and stress. So, people learn to take the drug to relieve these negative feelings. The result is that negative reinforcement teams up with positive reinforcement to "team-teach" drug self-administration. And this all happens unconsciously. Users are not aware of the lessons they are learning. Operant conditioning occurs each time a drug is used. The lesson it teaches — keep taking drugs — is therefore well learned.

How Tolerance, Physical Dependence, and Psychological Dependence Interact With Learning

As drug users consume more of a drug, they develop tolerance. To overcome tolerance, they increase the dose of the drug. Gradually, as they take higher and higher doses and use the drug more frequently, they discover that they need the drug just to feel normal. At this point, they have become physically dependent. Their brains and bodies have adapted to the drug and cannot function "normally" without it. If these drug users stop drug use, even for a short period of time, withdrawal symptoms appear. The more intense withdrawal becomes, the worse users feel. But, as soon as they take the next dose, they feel better.

The relief from withdrawal symptoms produces an especially potent form of negative reinforcement. So, having become physically dependent on a drug by abusing it, the user creates a situation in which he or she must then take that drug to prevent or eliminate even worse stresses — the anxiety, the flu-like symptoms of physical discomfort, and the cravings — that begin as soon as the last dose of a drug has worn off. This negative reinforcement becomes an especially important mechanism in maintaining addiction in people who use opiates, alcohol, and depressants.

As drug users integrate their drug use ever more deeply into their lives, they eventually develop psychological dependence, believing they can no longer live without the drug. They *have* to have their drug. When this happens, the drug user has become a drug addict.

Classical Conditioning

Addiction involves still another kind of learning — classical conditioning. Drug addicts learn to associate any number of neutral stimuli with the reward induced by drug-taking. These stimuli can include almost anything. For example, the sight of drug paraphernalia quickly elicits craving in abstinent addicts. Moreover, places (the car), activities (finishing dinner), simple items (a $10 bill), and people (the drug dealer) can have the same effect. Pretty soon, simply through the associations produced by classical conditioning, these stimuli themselves elicit measurable biological responses and drug craving.

Neutral stimuli that elicit drug craving are called "cues" or "triggers." For Henry, the process of shooting up began when he first received a glassine envelope containing a glistening white powder from a dealer. After many years, the simple sight of the glassine envelope, and then the powder, including its smell, became classically conditioned cues for Henry. While he was using heroin, Henry became excited as soon as he saw them, and he rushed to dissolve the powder, draw it into his syringe, and shoot it into a vein. Henry got excited because, to him, that white powder meant that he was about to get high. Even months after entering treatment, the sight of any white powder still makes Henry want to get high. He doesn't know why, but for him, white powder, such as confectioner's sugar, still makes Henry want to get high.

The power of classical conditioning is hard to underestimate. A drug addict like Henry — or Sybil, who smoked several dozen cigarettes each day — will have hundreds to thousands of trials associating drug use with other, formerly neutral, stimuli. Although they can be muted, these associations can never really be unlearned. They are always lurking, creating the risk of relapse, with the power to elicit craving even years after an addict has successfully withdrawn from the drug.

Explicit Learning and Memory

Still another kind of learning is involved in addiction. It is the conscious, explicit learning that produces most of our conscious memories. Drug users remember the pleasure and relief produced by drugs, and those prized memories can also produce a longing to use drugs again.

In many ways, explicit learning is the easiest kind of drug-related learning to overcome. The memories created by explicit learning are stored in the cerebral cortex and are available to consciousness. Memories created by either operant or classical conditioning remain outside of, and unavailable, to consciousness. But drug users can bring explicit memories directly into conscious aswareness to analyze and compare them with other pertinent information.

People can contrast a pleasant memory of drug use with information about the not-so-pleasant consequences of drug use. A memory of drinking some nice wine in a favorite restaurant, which may lead a recovering alcoholic to consider drinking again, can be directly overridden by the knowledge (another memory) that drinking produces horrible consequences for this particular person. He or she can then use reason and make a rational decision not to drink.

DRUGS TAKE BEHAVIORAL CONTROL AWAY FROM THE CEREBRAL CORTEX

Because drug addicts retain their ability to compare the risks and benefits of drug use, most people find it hard to understand why addicts don't "just say no." The logic is unassailable: drug use is bad for them. It makes perfect sense not to use drugs. Why don't they just stop? The reason is that the explicit learning that takes place consciously is only a small aspect of all the learning that takes place as addiction develops. More important, *rational thought cannot easily override all the unconscious, conditioned responses to drugs, which are acquired along with the addiction.*

Unconsciously learned responses are powerful because they arise in primitive parts of the brain (including the brain reward system) that have more direct control over human behavior than the cerebral cortex. The cortex, the site of reasoned evaluation of information and rational decision-making, can overcome motivations arising from more primitive brain regions only after much training and practice. Drugs act to usurp the normal role of the cerebral cortex in controlling behavior.

ADDICTION IS A BRAIN DISORDER

There is no simple explanation for drug addiction. It is a disorder formed by the confluence of many factors. The necessary event, the one that absolutely must occur, is *the repeated exposure of the brain to drugs that activate the brain reward system.* And, although the repeated exposure to addictive drugs is necessary for addiction to occur, it is not sufficient to cause addiction by itself. Addictive drugs have to be taken within a specific motivational context. That is, they have to be taken purposefully to produce pleasure or to reduce stress, discomfort, and anxiety. They have to be taken to get high. Somehow, this combination of drug and behavior allows drugs to change the brain, so that it begins to behave as if getting drugs were as important as getting food or drink. The addicted brain treats drug-taking as a survival function.

TREATMENT IS A TRANSITION

The long-term abuse of drugs causes profound changes in the brain. The behavior of addicts is strongly influenced by the maladaptive learning that takes place as addiction develops. As a result, recovering from drug addiction does not mean returning to a condition like the one that existed before drug abuse began. Instead, addicts must grow into a new level of personal awareness, with new patterns of behavior. That is one reason why the treatment of addiction is so difficult. Still, treatment is effective, and the benefits to the addict, his family, and society are quite large.

Treatment is not easy and quick. More commonly, it is characterized by one or more episodes of relapse into uncontrolled drug use before the former addict really gains control of his or her behavior. Recognizing the chronic relapsing nature of addiction, however, allows us to view it, and its treatment, in a new light. Just as a detour in a long car ride does not mean the driver will never reach his destination, neither does relapse into drug addiction have to mean that treatment has failed. Remaining engaged in the process is important, even while it appears that the outcome may be in doubt. Like any other aspect of life that requires new learning, the learning of the new behaviors that will allow an addict to live without abusing drugs takes time. We need to allow recovering addicts to be occasionally fallible, just like the rest of us, as they work to undo a great harm they have done themselves.

NEUROSCIENTISTS SEEK WAYS TO TREAT THE BRAIN

If drug addiction is a brain disorder, one of the goals of treatment research should be to seek ways to treat the brain directly. A crucial first step in the process is to continue to search for the changes in the brain that underlie addiction. As such knowledge has begun to accumulate, neuroscientists are turning their attention to developing medications that will treat the brain as part of a comprehensive treatment program. But an even more profound area of investigation may lie beyond, as we will discuss in our concluding pages.

Epilogue

Neurologists have always viewed changes in behavior as good indicators that something in the brain has changed. So, when we observe that drug addiction involves profound behavioral changes, we do not have to make a large logical leap to interpret this as evidence that something about the brain has actually changed. Indeed, if drug abuse were not historically so entwined in complex legal, moral, and social issues, an observer without any knowledge of these contingencies — like a visitor from Mars — would simply conclude that drug addiction is a brain disorder.

Despite the simple logic of our visitor from Mars, most people find it difficult to accept addiction as a brain disorder. Several reasons contribute to this reluctance. (1) Because the public is generally unaware of the advances neuroscience has made, few people understand how, or even that, drugs change the brain. (2) Most find it difficult to comprehend how something that begins as a voluntary behavior can end up as an involuntary compulsion. (3) People have historically viewed the mind and the brain as separate. (4) The idea that humans have free will has been such an important foundation of Western moral philosophy and religion that it is virtually inconceivable that anything could take free will away. (5) Drug addiction is still commonly seen as a moral failing that can be corrected if the addict simply will exert a little more "will power."

Lacking information about how drugs change the brain, believing that the mind and brain are somehow separate, and believing that our free will resides only in our mind (it does seem that way) have made it easier to regard drug addiction as little more than the loss of will power, a condition the addict could correct if he or she simply cared enough to do something about it.

Actually, it is still appropriate to interpret the loss of control over drug-taking as a loss of will power, which, at the very least, represents an impairment of free will. What is different now is that we ascribe this impairment of free will not to a moral failure, but to a brain disorder. By embracing the idea that the brain and the mind are a single entity — indeed that the brain *is* the mind — we can see that *free will resides in the brain*.

201

Addictive drugs somehow change parts of the brain that influence our ability to exercise free will and, once a person is addicted, these changes override our freedom to choose how to behave. Understanding this presents a stunning opportunity. If free will resides in the brain, and we can determine (1) which parts of the brain are altered by addiction (resulting in loss of free will), (2) how those parts of the brain work normally, and (3) how addiction changes the way those parts function, we will be on the trail to understanding how the brain makes decisions. In other words, we have an opportunity to discover the neural basis of free will.

Thus, understanding how the brain works is all about understanding what it means to be human and neuroscience stands at the interface between our quest to understand the physical world and our desire to understand ourselves.

In the end, the effort to study and understand addiction may not only result in ways to treat addiction more effectively. It may also open a door into one of the most fundamental aspects of human nature, our ability to chose consciously what we will do, and thereby become.

Glossary

Absorption   The process the body uses to move elements from the outside world into the blood and other tissues. Food is absorbed through the stomach and intestines. Nicotine is absorbed through the lungs.

Abstinence   The conscious choice not to use drugs. The term "abstinence" usually refers to the decision to end the use of a drug as part of the process of recovery from addiction.

Acetaldehyde   The metabolite that results when alcohol dehydrogenase breaks down alcohol in the body.

Acetylcholine   A neurotransmitter. Acetylcholine is used by spinal cord neurons to control muscles and by many other neurons in the brain as well. Nicotine binds to one type of acetylcholine receptor.

Action potential   The electrical part of a neuron's two-part, electrical-chemical message. An action potential consists of a brief pulse of electrical current that travels along the axon to relay messages over long distances.

Acute effects   The short-term effects of a drug. Acute effects are those that people feel shortly after they ingest a drug and are under its influence (e.g., while they are intoxicated).

Adaptive behaviors   Useful behaviors we acquire as we respond to the world around us. Adaptive behaviors help us get the things we want and need for life.

Addiction   A brain disorder characterized by the loss of control of drug-taking behavior, despite adverse health, social, or legal consequences to continued drug use. Addiction tends to be chronic and to be characterized by relapses during recovery.

Addictive drugs   Drugs that change the brain, change behavior, and lead to the loss of control of drug-taking behavior.

Adenosine   A neurotransmitter that binds to the adenosine receptor. Caffeine is an adenosine antagonist and prevents adenosine from binding with its receptor.

Adrenal gland   A small gland in the body that releases a variety of hormones that help us deal with stress. Two of these hormones, epinephrine and norepinephrine, are also part of the flight-or-fight response. Cocaine sharply increases the levels of these hormones in the body.

Agonist   A chemical that binds to a specific receptor and produces a response, such as excitation or inhibition of action potentials. Opiates, cannabis, nicotine, and some hallucinogens such as LSD are agonists.

Alcohol   A chemically simple, but psychoactively complex drug commonly used in many beverages. Alcohol is a depressant drug with significant liability for abuse and addiction.

Alcohol dehydrogenase   The enzyme found mainly in the liver and stomach that breaks down (metabolizes) alcohol.

Alcoholics Anonymous   One of the earliest forms of addiction treatment in the United States, AA developed the 12-step approach to assisting recovery from alcohol addiction (alcoholism). Several other anonymous groups have adapted the 12-step approach to help people recover from addiction to other drugs (e.g., Narcotics Anonymous, Cocaine Anonymous, Pot Smokers Anonymous).

Alveoli   Tiny, balloon-like air sacks in the lungs. Alveoli are designed to allow oxygen to pass rapidly into the blood and are also efficient at absorbing inhaled drugs.

Alzheimer's disease   A degenerative disease in which neurons of the brain die, leading to the loss of the ability to think, learn, and remember (dementia).

Amino acids   Small chemical compounds that are the building blocks of proteins.

Amphetamines   Stimulant drugs whose effects are very similar to cocaine.

Analgesics   Drugs that relieve pain.

Analogs   Drugs whose chemical structures have been slightly modified from a parent compound. There are many analogs to morphine or to LSD. See *Designer drug*.

Anandamide   The endogenous neurotransmitter that binds to the cannabinoid receptor.

Anesthesia   The loss of sensation, primarily to pain, often accompanied by the loss of consciousness.

Anesthetic gases   Gaseous drugs that produce loss of sensation and consciousness.

Antagonist   A chemical that binds to a receptor and blocks it, producing no response, and preventing agonists from binding, or attaching, to the receptor. Antagonists include caffeine and naloxone.

Assessment   The diagnostic process in which a professional examines a drug user to determine the extent of the person's drug use, whether he or she is addicted, and what type of treatment might be most effective.

Auditory cortex   That part of the cerebral cortex that processes sounds and produces our awareness of them.

Axon   The cable-like structure neurons used to send messages to other neurons. It carries the neuron's electrical message.

Axon terminal   The structure at the end of an axon that produces and releases chemicals (neurotransmitters) to transmit the neuron's message across the synapse to another neuron.

Barbiturates   Depressant drugs that produce relaxation and sleep. Barbiturates include sleeping pills such as pentobarbital (Nembutal) and secobarbital (Seconal).

Basal ganglia   The large, complex set of brain structures involved in generating movements, in some cognitive functions, and in some emotional and motivational activities. The basal ganglia and the cerebral cortex work together to refine movements, thoughts, and feelings.

Behavior   The observable activity of humans and animals.

Behaviorism   The study of behavior, especially using operant conditioning.

Benzodiazepines   The so-called "minor" tranquilizers, depressants, which relieve anxiety and produce sleep. Benzodiazepines include tranquilizers such as diazepam (Valium) and alprazolam (Xanax) and sleeping pills such as flurazepam (Dalmane) and triazolam (Halcion).

Bernard, Claude   The physiologist who coined the term "homeostasis."

Bind   What occurs when a neurotransmitter attaches itself to a receptor. The neurotransmitter is said to "bind" to the receptor.

Binge   Uninterrupted consumption of a drug for several hours or days.

Bolus   A concentrated amount of drug; a dose injected rapidly into a vein, a rounded mass of matter.

Brain   That part of the central nervous system inside our heads. Our brain is the seat of all our perceptions, thoughts, feelings, and voluntary movements.

Brain reward system   A brain circuit that, when activated, reinforces behaviors. The circuit includes the dopamine-containing neurons of the ventral tegmental area, the nucleus accumbens, and part of the prefrontal cortex. We perceive the activation of this circuit as pleasure.

Brain stem   The relatively primitive brain structure that starts where our spinal cord enters our head. Neurons within the brain stem control basic functions such as heart rate and breathing.

Broca, Paul   The scientist who identified the area in the brain responsible for producing speech, now called *Broca's area*.

Buprenorphine   A long-lasting opiate analgesic that has both opiate agonist and antagonist properties. Buprenorphine shows promise for treating heroin addiction.

Caffeine   A mild stimulant, the most widely used drug in the world.

Cannabinoid receptor   The receptor in the brain that recognizes THC, the active ingredient in marijuana. Marijuana exerts its psychoactive effects via this receptor.

Cannabis   The botanical name for the plant from which marijuana comes.

Capillaries   The smallest blood vessels. Oxygen and nutrients leave the bloodstream through capillaries to get into the body. Gases from the alveoli enter the bloodstream through capillaries in the lungs.

Cell body   The central structure of a neuron, which contains all of the molecular parts that keep the cell alive, generate new parts, and repair or destroy existing parts.

Cell membrane   The outside covering, or "skin" of a cell. Receptors and ion channels are embedded in it.

Cellular metabolism   The production of energy and new materials in a cell.

Central nervous system   The brain and spinal cord.

Cerebral cortex   The large, deeply folded outer layers of the brain that make our heads so big. The cortex carries out complex perceptual, cognitive, and motor tasks.

"China white"   A "designer drug" that was an opiate derivative. Some batches contained a neurotoxin called MPTP, which killed neurons that make dopamine, producing symptoms similar to Parkinson's disease.

Cholinergic   The adjective derived from acetylcholine. A neuron that contains acetylcholine is a cholinergic neuron.

Circuits   A group of cortical fields or nuclei that are linked together by their axons to perform a specific brain function. Core components of circuits are constantly in touch with each other, whereas other components can be brought in as the need arises.

Classical conditioning   The form of implicit, unconscious learning in which a neutral stimulus becomes associated with a significant stimulus through repeated pairing of the two.

Cocaine   A highly addictive stimulant drug derived from the cocoa plant that produces profound feelings of pleasure. See *Crack*.

Codeine   A natural opioid compound that is a relatively weak, but still effective, opiate analgesic. It has also been used to treat other problems (e.g., to relieve coughing).

Cognitive functions   Higher brain functions involving the manipulation of information from the senses and from memory. They often require awareness and judgment, and they enable us to know and to analyze problems and plan solutions — in short, to think.

Consciousness   Our own awareness of ourselves and the world; the mental processes that we can perceive; our thoughts and feelings.

Cortical field   A large aggregation of millions of nerve cells in a circumscribed region of the cerebral cortex, which together carry out a specific function, receive connections from the same places, and have a common structural arrangement. There are many dozens of such fields in the cerebral cortex. Elsewhere in the brain such groups are called *nuclei*.

Crack   A chemically altered form of cocaine that is smoked.

Craving   Hunger for drugs. It is caused by drug-induced changes that occur in the brain with the development of addiction and arises from a need of the brain to maintain a state of homeostasis that includes the presence of the drug.

Cues   Formerly neutral stimuli that acquire the ability to elicit drug-craving through classical conditioning. Cues are also called *triggers*.

Dalmane   A depressant drug of the benzodiazepine family used to produce sleep.

Decondition   The unlearning of classically conditioned responses. Helping addicts identify and neutralize the cues or triggers they developed while they were addicted.

Dendrites   The branches that reach out from a neuron's cell body to receive messages from the axon terminals of other neurons.

Denial   Unconsciously refusing to admit that someone is addicted. Denial occurs among addicts themselves and among those who are close to them.

Dentate gyrus   A key part of the hippocampus that contains one of the highest concentrations of cannabinoid receptors in the brain.

Deoxyribonucleic acid (DNA)   The chemical compound that makes up genes.

Depressants   Drugs that relieve anxiety and produce sleep. Depressants include barbiturates, benzodiazepines, and alcohol.

Designer drug   An illegally manufactured chemical whose molecular structure is altered sightly from a parent compound to enhance specific effects. Examples include DMT, DMA, DOM, MDA, and MDMA (ecstasy).

Detoxification   The process of removing a drug from the body. This is the initial period addicts must go through to become drug-free. Withdrawal symptoms appear early during this process. Depending on the drug, detoxification lasts for a few days to a week or more.

Diversion   Taking legally prescribed medications (e.g., methadone, tranquilizers) and selling them illegally.

DMA   A hallucinogenic "designer drug" with psychedelic properties.

DMT   A hallucinogenic "designer drug" with psychedelic properties.

DOM   A hallucinogenic "designer drug" with psychedelic properties.

Dopamine   The neurotransmitter that produces feelings of pleasure when released by the brain reward system.

Dopamine transporter   A structure that straddles the cell membranes of axon terminals of dopamine-releasing neurons and rapidly removes dopamine from the synapse.

Double-blind trials   Studies of an experimental drug in which neither patient nor doctor knows whether the patient is receiving the experimental drug or some alternative (which might be a placebo if no treatment already exists).

Dronabinol   The generic name of synthetic THC.

Drug abuse   Using illegal drugs; using legal drugs inappropriately. The repeated, high-dose, self-administration of drugs to produce pleasure, to alleviate stress, or to alter or avoid reality (or all three).

Drug addiction   See *Addiction*.

Drug-free treatment   An approach to helping addicts recover from addiction without the use of medications.

Drug treatment   A combination of detoxification, psychosocial therapy and, if required, skill acquisition to help people recover from addiction.

Dynorphins   Peptides with opiate-like effects that are made by neurons and used as neurotransmitters; one of the endogenous opioids that binds to opiate receptors.

Ecstasy (MDMA)   A chemically modified amphetamine that has hallucinogenic as well as stimulant properties.

Enabling   Things that people who are close to addicts do unconsciously that either encourage, or at least do not interfere with, the addict's drug use.

Endogenous   Something produced by the brain or body.

Endorphins   Peptides with opiate-like effects that are made by neurons and used as neurotransmitters; one of the endogenous opioids that binds to opiate receptors.

Enkephalins   Peptides with opiate-like effects that are made by neurons and used as neurotransmitters; one of the endogenous opioids that binds to opiate receptors.

Enzyme   A large molecule that living organisms use to facilitate the transition from one form of a chemical to another. Enzymes are used to build, modify, or break down different molecules.

Ether   An inhalant. Ether was one of the first anesthetics to be used in surgery, but has been replaced by more effective, safer anesthetics.

Euphoria Intense pleasure. Drug-induced euphoria is a "rush" of pleasurable feelings. It is caused by the release of the neurotransmitter, dopamine, within the brain reward system.

Excitatory neurotransmitter A neurotransmitter that acts to elicit an action potential or make it more likely that one will be elicited.

Explicit memory Memories derived from conscious learning, using our senses and attention to store information about what is in the world and where and when events have occurred.

Fight-or-flight response An automatic response of our body that prepares us to act to save ourselves when we become excited or scared.

Free will Our ability to make choices and decisions that are not under the control of outside forces or prior causes.

GABA (gamma-aminobutyric acid) The major inhibitory neurotransmitter in the brain.

Gene Strands of DNA that contain the blueprint of all the molecules that make up our bodies.

Glial cells Tiny brain cells that support neurons by performing a variety of "housekeeping" functions in the brain.

Glucose A simple sugar that the brain uses as its major source of energy.

Glutamate The most common excitatory neurotransmitter in the brain.

Habilitate The process of teaching the skills needed for successful living. Habilitation helps people recover from addiction by teaching life skills that were never learned because drug use interfered with the learning and maturation process. Habilitation is especially important for addicts who started drug use young.

Halcion A depressant drug of the benzodiazepine family used to induce sleep.

Hallucinogens A diverse group of drugs that alter perceptions, thoughts and feelings. Hallucinogens do not produce hallucinations. These drugs include LSD, mescaline, MDMA (ecstasy), PCP, and psilocybin (magic mushrooms).

Heroin   The potent, widely abused opiate that produces a profound addiction. It consists of two morphine molecules linked together chemically.

Hippocampus   A brain structure that is involved in emotions, motivation, and learning. It plays an important role for short-term (working) memory and is crucial for our ability to form long-term memories.

Homeostasis   The process of keeping the internal environment of the body stable while the outside world changes.

Hypothalamus   The part of the brain that controls many bodily functions, including feeding, drinking, and the release of many hormones.

Implicit memory   The memories acquired through unconscious learning processes, such as operant and classical conditioning.

Inhalants   Any drug administered by breathing in its vapors. Most inhalants are organic solvents such as glue and paint thinner, or anesthetic gases such as ether and nitrous oxide.

Inhibitory neurotransmitter   A neurotransmitter that acts to prevent a neuron from firing an action potential.

Inpatient treatment   Residential treatment for drug addiction in a hospital or clinic.

Interneuron   Any neuron that only sends its messages locally (within a millimeter or so). Many are inhibitory.

Intervention   The act of interrupting addiction and persuading the addict to enter treatment.

Intervention counselor   A person who conducts an intervention with an addict and the addict's family and close friends.

Intoxication   Being under the influence of, and responding to, the acute effects of a psychoactive drug. Intoxication typically includes feelings of pleasure, altered emotional responsiveness, altered perception, and impaired judgment and performance.

Kinesthetic information Information from our muscles and joints that tells us where our body is in space and how its various parts are oriented in relation to each other. Kinesthetic information is crucial for making accurate movements.

Korsakoff's syndrome See *Wernike-Korsakoff's syndrome*.

LAAM A very long-lasting opiate agonist recently approved for the treatment of opiate addiction.

Ligand Any chemical that binds to a receptor. Ligands may be agonists or antagonists.

Limbic system A set of brain structures that generates our feelings, emotions, and motivations. It is also important in learning and memory.

Localization of function A principal of brain organization that states that specific places (circuits) in the brain carry out specific functions.

Locus coeruleus A group of neurons (nucleus) that is the source of all of the neurotransmitter norepinephrine in the brain.

Long-term effects The effects seen when a drug is used repeatedly over weeks, months, or years. These effects may outlast drug use.

Long-term memory Enduring memories about things, places, and events.

Long-term memory circuit The brain circuit, including the cerebral cortex and hippocampus, which enables the brain to lay down and store memories in the cortex.

LSD An hallucinogenic drug that acts on the serotonin receptor.

Maintenance treatment Treatment for opiate addiction that involves giving the addict a synthetic opiate (methadone or LAAM) to prevent the withdrawal and craving that often provoke relapse.

Maladaptive behaviors Behaviors acquired by drug users that hinder them from succeeding in the normal, non-drug-using world.

Marijuana   A psychoactive drug made from the leaves of the cannabis plant. It is usually smoked but can also be eaten. See *Cannabis*.

Marinol   The trade name of dronabinol, a synthetic version of THC used as medicine.

MDA   One of several hallucinogenic "designer drugs" with psychedelic properties that are manufactured by basement chemists.

MDMA (Ecstasy)   A hallucinogenic "designer drug" with psychedelic and stimulant properties.

Mescaline   A naturally occurring hallucinogenic drug that acts on the serotonin receptor.

Messenger ribonucleic acid (mRNA)   A molecule that carries the genetic code from DNA to the parts of the cell that use the code to make components of the cell.

Metabolic enzymes   Enzymes that break down or inactivate drugs in the body; also, enzymes that break down food and produce energy.

Metabolic tolerance   The body's increased ability to eliminate a drug, thereby making a given dose less effective.

Metabolism   The processes by which the body breaks things down or alters them so they can be eliminated; also, the processes by which the body extracts energy and nutrients from food.

Metabolites   The products that result when enzymes in the body break things down or alter them to produce energy or eliminate them.

Methadone   A long-lasting synthetic opiate used to treat cancer pain and heroin addiction.

Methamphetamine   A commonly abused, potent stimulant drug that is part of a larger family of amphetamines.

Microsomal ethanol oxidizing system (MEOS)   Liver enzymes that metabolize many drugs, including alcohol.

Mind The container of the contents of consciousness, what we call the results of our processes of perception, thinking, and feeling. The mind is the manifestation of consciousness.

Morphine The most potent natural opiate compound produced by the opium poppy.

Motivation The internally generated state (feeling) that stimulates us to act. The neural substrate for motivation is most likely found in the brain reward system.

Motor cortex The part of the cerebral cortex that creates the commands that make the muscles move.

Motor neurons The neurons that control our muscles.

MPTP A neurotoxin, found in a "designer" opiate called "China white," which kills the neurons that make dopamine, producing a set of symptoms that look like Parkinson's disease.

Myelin sheath A covering made of a special fat that encloses a neuron's axon and allows it to transmit action potentials.

Naloxone A short-acting opiate antagonist that binds to opiate receptors and blocks them, preventing opiates from binding to these receptors. Naloxone is used to treat opiate overdoses.

Naltrexone A long-lasting opiate antagonist used for the treatment of heroin addiction, and more recently used for the treatment of alcohol addiction.

Negative reinforcement Reward generated by the removal of painful or stressful conditions or events.

Nembutal (pentobarbital) A depressant drug of the barbiturate family used to induce sleep.

Neural substrate The set of brain structures that underlies specific behaviors or psychological states.

Neurochemicals Neurotransmitters and other brain chemicals produced by neurons.

Neuron   Nerve cell. Neurons are unique cells found in the brain and body that are specialized to process and transmit information.

Neuroscience   The study of how the brain and nervous system work. Neuroscience integrates more traditional scientific approaches such as anatomy, physiology, and biochemistry, along with newer fields such as molecular biology and computer science, to understand how the nervous system functions.

Neurotoxins   Substances that damage or kill neurons.

Neurotransmission   The process that occurs when a neuron releases neurotransmitters to communicate with another neuron across the synapse.

Neurotransmitter   Chemicals produced by neurons to carry their messages to other neurons.

Nicotine   The drug in tobacco that is addictive. Nicotine also activates a specific kind of acetylcholine receptor.

Nicotine gum, nicotine patch   Two methods of delivering small amounts of nicotine into the bodies of people who are addicted to nicotine to help them quit smoking cigarettes by preventing nicotine withdrawal.

Nicotinic cholinergic receptor   One of two acetylcholine receptors. This one responds to nicotine as well as acetylcholine.

Nitrous oxide   An inhalant, also known as "laughing gas." Nitrous oxide is a weak anesthetic that does not produce unconsciousness.

Norepinephrine   A neurotransmitter and a hormone. It is released by the sympathetic nervous system onto the heart, blood vessels, and other organs and by the adrenal gland into the bloodstream as part of the fight-or-flight response. Norepinephrine is also present in the brain and is used as a neurotransmitter in normal brain processes.

Nucleus   A cluster or group of nerve cells that is dedicated to performing its own special function(s). Nuclei are found in all parts of the brain except the cerebral cortex, where such groups are called *cortical fields*.

Nucleus accumbens   A part of the brain reward system, located in the limbic system, that processes information related to motivation and reward.

It is the key brain site where virtually all drugs of abuse act to reinforce drug taking.

Open-label study   A study in which both doctor and patient know that patients are receiving an experimental drug and what that drug is.

Operant conditioning   An unconscious form of learning in which a behavior is linked to a specific stimulus through a process of reinforcement.

Opiate receptors   Receptors that recognize both opiates and endogenous opioids. When activated, they slow down or inhibit the activity of neurons on which they reside.

Opiates   Any of the psychoactive drugs that originate from the opium poppy or that have a chemical structure like the drugs derived from opium. Such drugs include opium, codeine, and morphine (derived from the plant), and hydromorphone (Dilaudid), methadone, and meperidine (Demerol), which were first synthesized by chemists.

Opioid   Any chemical that has opiate-like effects; commonly used to refer to endogenous neurochemicals that activate opiate receptors.

Organic solvents   One class of inhalants that includes substances such as gasoline, paint thinner, and glue. Organic solvents are neurotoxic because they dissolve fatty substances, including the axon's myelin sheath.

Outpatient treatment   Nonresidential treatment for drug addiction. Patients live at home, often work, and come to a clinic for treatment.

Overdose   The condition that results when too much of a drug is taken, making a person sick or unconscious and sometimes resulting in death.

Parallel processing   When various cortical fields and nuclei work together simultaneously, each on a small part of a bigger information-processing job.

Paranoid schizophrenia   A severe form of mental illness typically characterized by delusions of persecution and hallucinations. This condition may be induced by binge use of stimulants.

Parkinson's disease   A disease in which dopamine-containing neurons die. It produces severe impairments in movement, cognitive function, and emotions.

PCP (phencyclidine)   PCP has an array of effects. Originally developed as an anesthetic, it may act as an hallucinogen, stimulant, or sedative.

Peptides   Small protein-like compounds made of amino acid building blocks.

Perception   The conscious awareness of sensory inputs, internal states, or memories.

Periaqueductal gray matter   A set of nuclei deep within the brain stem that are involved with visceral functions. It also plays a role in the development of physical dependence on opiates.

Pharmacodynamics   The study of the mechanisms of actions of a drug, the relationship between how much drug is in the body and its effects.

Pharmacokinetics   The study of how the body absorbs drugs, how they are distributed throughout the body, and how the body gets rid of drugs.

Phencyclidine   See *PCP*.

Physical dependence   Changes that the brain and body undergo as they adapt to the continued presence of high doses of drugs. Because of these changes, the brain and body eventually come to require the presence of the drug to work properly.

Placebo   An inactive substance.

Plasticity   The capacity of the brain to change its structure and function within certain limits. Plasticity underlies brain functions such as learning and allows the brain to generate normal, healthy responses to long-lasting environmental changes.

Positive reinforcement   Something that increases the likelihood that the behavior that elicited it will be repeated. Positive reinforcement is rewarding, and we typically perceive it as pleasure.

Positron emission tomography (PET)   A technique for measuring brain function in living human subjects by detecting the location and concentration of tiny amounts of radioactive chemicals.

PET scanner   The machine that detects the radioactive chemicals used to measure brain function.

Postsynaptic neuron   A neuron that receives messages from neurons on the other sides of its synapses.

Prefrontal cortex   The part of the cerebral cortex at the very front of the brain. It is involved with higher cognitive and emotional functions including short-term memory, learning, and setting priorities for future actions.

Presynaptic neuron   A neuron that releases neurotransmitters into synapses to send messages to other neurons.

Prevention   Stopping drug use before it starts, intervening to halt the progression of drug use once it has begun, changing environmental conditions that encourage addictive drug use.

Primary reinforcers   Stimuli, such as food and water, which produce reward directly, with no learning about their significance or other intervening steps required. Most drugs of abuse are primary reinforcers.

Projection neurons   Neurons (usually excitatory) that send their axons away from the local vicinity to communicate with other parts of the brain.

Proteins   Large molecules made up of amino acid building blocks.

Psilocybin   A natural hallucinogenic drug derived from a mushroom. It acts on the serotonin receptor.

Psychedelic drug   Drugs that distort perception, thought, and feeling. This term is typically used to refer to drugs with actions like those of LSD.

Psychoactive drug   A drug that changes the way the brain works.

Psychological dependence   When drugs become so central to a user's life that the user believes he must use them.

Psychosis   Severe mental illnesses characterized by loss of contact with reality. Schizophrenia and severe depression are psychoses.

Psychosocial therapy   Therapy designed to help addicts by using a combination of individual psychotherapy and group (social) therapy approaches to rehabilitate or provide the interpersonal and intrapersonal skills needed to live without drugs.

Receptor   A large molecule that recognizes specific chemicals (normally neurotransmitters, hormones, and similar endogenous substances) and transmits the message carried by the chemical into the cell on which the receptor resides.

Rehabilitate   Helping a person recover from drug addiction. Rehabilitation teaches the addict new behaviors to live life without drugs.

Relapse   In general, to fall back to a former condition. Here, resuming the use of a drug one has tried to stop using. Relapse is a common occurrence in many chronic disorders that require behavioral adjustments to treat effectively.

Respiratory center   A small set of nuclei in the brain stem that regulate the speed and depth of breathing ultimately by controlling the muscles that move our chest and diaphragm.

Reuptake   The process by which neurotransmitters are removed from the synapse by being "pumped" back into the axon terminals that first released them.

Reuptake pump   The large molecule that actually transports neurotransmitter molecules back into the axon terminals that released them.

Reward   The process that reinforces behavior. It is mediated at least in part by the release of dopamine into the nucleus accumbens. Human subjects report that reward is associated with feelings of pleasure.

Rock   A small amount of crack cocaine in a solid form; free-base cocaine in solid form.

Route of administration   The way a drug is put into the body. Eating, drinking, inhaling, injecting, snorting, smoking, and absorbing a drug through mucous membranes all are routes of administration used to consume drugs of abuse.

"Run"   A binge of (more or less) uninterrupted consumption of a drug for several hours or days. This pattern of drug use is typically associated with stimulants, but is seen with alcohol as well.

Rush   Intense feelings of euphoria a drug produces when it is first consumed. Drug users who inject or smoke drugs describe their rush as being sometimes as intense, or even more intense, than sexual orgasm.

Seconal   A depressant drug of the barbiturate family that induces sleep.

Second messenger   A molecule produced inside neurons as a step in the process of communication between cells. The second messenger lets other parts of the cell know that a specific receptor has been activated, thereby completing the message carried by the neurotransmitter that bound to the receptor. Some receptors (e.g., dopamine, opiate) use second messengers. Others (e.g., nicotine, GABA) do not.

Secondary reinforcers   Formerly neutral stimuli that acquire the ability to produce reward through the learned association with a primary reinforcer. Money and praise are secondary reinforcers.

Sensitization   An increased response to a drug caused by repeated administration. Sensitization is most commonly seen in some responses to stimulants.

Serotonin   A neurotransmitter involved in many functions, including mood, appetite, and sensory perception.

Short-term effects   The acute effects of a drug. The effects felt during and shortly after the time a person is under the influence of (intoxicated by) a drug.

Short-term memory   Also called "working memory," short-term memory enables us to use information from our senses and from our memory and hold that information in our consciousness long enough to think about it.

"Skinner Box"   A device that automatically released food in response to an animal manipulating a specific object (e.g., pressing a bar). This device allowed scientists to measure behavior accurately over long periods of time.

Skin popping   Injecting a drug under the skin.

Somatosensory cortex   A brain region that processes information coming from the muscles, joints, and skin.

Stimulants   A class of drugs that elevates mood, increases feelings of well-being, and increases energy and alertness. These drugs also produce euphoria and are powerfully rewarding. Stimulants include cocaine, methamphetamine, and methylphenidate (Ritalin).

Stimulus   Any object or action that penetrates awareness or excites an animal to respond.

Stroke   The blockade or rupture of a blood vessel in the brain. This prevents oxygen from reaching neurons and may result in their death.

Structural proteins   Special proteins that form a framework for the cell bodies, dendrites, and axons of neurons.

Synapse   The site where neurons communicate with each other. A synapse is a small gap that physically separates neurons. Axon terminals of a neuron sending a message (the presynaptic neuron) release neurotransmitters into the synapse. The neurotransmitters diffuse to the other side (the postsynaptic side) where they bind to receptors on the postsynaptic neurons, thereby relaying the message.

Synaptic transmission   See *Neurotransmission.*

Synthesize   To make a chemical from constituent parts. Exact copies of drugs found in nature or created in the laboratory are synthesized in laboratories from simpler chemicals. Many substances are also synthesized in cells (e.g., large proteins such as receptors, or smaller ones such as neurotransmitters).

Tetrahydrocannabinol (THC)   The major active ingredient in marijuana. It is primarily responsible for producing the high and the rest of the drug's psychoactive effects.

Thalamus   A brain structure that lies between the brain stem and the cortex and acts as a relay to the cortex for almost all sensory inputs and other kinds of information.

THC   See *Tetrahydrocannabinol.*

Theobromine   A mild stimulant found in tea and cocoa. It is a chemical cousin of caffeine.

Theophylline   A chemical cousin of caffeine that is found in tea.

Therapeutic communities Communities that provide long-term, residential treatment for drug addiction, offering detoxification, group therapy, and skill acquisition.

Titrate Adjust the dose of a drug to a desired level.

Tolerance A physiological change resulting from repeated drug use that requires the user to consume increasing amounts of the drug to get the same effect a smaller dose used to give.

Toxic Poisonous.

Tranquilizers Depressant drugs that relieve anxiety.

Transdermal absorption Absorption through the skin.

Transporter A large molecule that straddles the cell membrane of the axon terminals of neurons. It removes neurotransmitter molecules from the synapse by ferrying them back into the axon terminal that released them.

Triggers Formerly neutral stimuli that have attained the ability to elicit drug craving following repeated pairing with drug use; also called *cues*.

Valium A depressant drug of the benzodiazepine family that relieves anxiety.

Ventral tegmental area (VTA) The group of dopamine-containing neurons that make up a key part of the brain reward system. The key targets of these neurons include the nucleus accumbens and the prefrontal cortex.

Vesicles Tiny sacks within axon terminals that produce, release, and store neurotransmitters.

Visual cortex A brain region in the back of the head that allows us to perceive the visual information gathered by our eyes.

Wernicke, Carl The scientist who discovered the area of the cerebral cortex that allows us to understand language. People with damage in this area of the cortex are unable to understand spoken or written words.

Wernicke-Korsakoff's syndrome A brain disorder characterized by the loss of the brain's ability to store memories.

Withdrawal   Physical symptoms in the body and brain that occur after cessation of drug use in a person who is physically dependent on that drug.

Working memory   See *Short-term memory*.

Xanax (*aprazolam*)   A depressant drug of the benzodiazepine family that relieves anxiety.

INDEX

12-Step Programs, 160–161

A

Absorption, 203
Abstinence, 203
Acetaldehyde, 119, 203
Acetylcholine, 60, 70, 203
  acetylcholine receptors, 60
Action Potential, 32–33, 35–38,
  42–48, 52, 57, 203
  effects of opiates, 54–55, 57,
  126; effects of nicotine, 60–61;
  effects of depressants, 64; effects
  of alcohol, 68; effects of
  stimulants, 69; effects of
  PCP, 75
Acute Effects, 203
Adaptive Behaviours, 203
Addiction, 116, 134–137,
  139–151, 153–173,
  178–181, 184–189,
  195–202, 203
Addictive Drugs, 204
Adenosine, 63, 204
  adenosine receptors, 63;
  effects of caffeine, 63
Adrenal Gland, 146, 204
Adrenaline, 6
Agonist, 51, 53, 55, 63, 204
  opiates, 53–57, 165–168;
  marijuana, 57–60; nicotine,
  60–61; hallucinogens, 61–62;
  LSD, 62; alcohol, 66–68

AIDS, 165, 181
  treatment with THC, 82, 177,
  189–190
Alcohol, 2, 8–10, 37, 49–51, 56,
  63, 78, 150, 192, 195, 204
  cause of memory loss, 9, 98, 110;
  damage to body organs, 9, 65, 69;
  effects on serotonin receptors, 62,
  66–69; role as depressant, 64–69;
  reduction of heart disease, 65;
  effects on breast cancer, 65;
  effects on GABA receptors,
  66–67; effects on glutamate
  receptors, 66–69; effects on the
  brain reward system, 66;
  effects on dopamine receptors, 66,
  68; effects on respiratory
  system, 68; effects on cell
  membranes, 68–69; effects on
  dopamine neurons, 75; route of
  administration, 83–84, 88;
  breakdown by metabolism, 89,
  119–120; role of conditioning,
  105; destroys neurons, 110;
  intoxication from, 116–117;
  tolerance to, 118–119, 123;
  sensitization to, 124–125;
  dependence on, 126–127;
  withdrawal from, 128; pattern of
  use, 142; testing for, 155;